W9-AFJ-103

The Courtship of Birds

The

Courtship of Birds

HILDA SIMON

Illustrated by the Author

DODD, MEAD & COMPANY · NEW YORK

Printed in the United States of America

1 2 3 4 5 6 7 8 9 10

Library of Congress Cataloging in Publication Data

Simon, Hilda.
The courtship of birds.

Bibliography: p.
Includes index.
1. Birds—Behavior. 2. Courtship of animals.
I. Title.
QL698.3.S53 598.2'5 77-3211
ISBN 0-396-07459-6

TO S. PHELPS PLATT, JR.
President, Dodd, Mead & Company
whose interest, understanding, and enthusiastic cooperation
were invaluable contributions to my work on this book

Contents

Illustrations

The Courtship of Birds

1. Patterns of Courtship Behavior

FROM ITS PERCH near the top of the tall old tree growing in the suburban backyard, the robin poured forth its simple but full-throated and melodious song. Wings drooping and feathers slightly fluffed, the bird known and loved throughout the American north-east as a harbinger of spring sat all but motionless near the tip of the broken branch extending beyond the tree's crown. From there it could survey its domain while informing the world that the time for mating and raising a family had once again arrived. Only a few days earlier, that same male's song had attracted and won a female who was now busy constructing a nest, while he sang not only to entertain and encourage his mate, but also to warn off potential rivals and reassert his claim to the territory he considered his own.

The ways in which birds woo and win their mates are almost as diverse as the variations in size, coloration, and general habits that distinguish the more than 8,500 species of this large animal class. Although it would be an exaggeration to state that the number of different courtship practices matches the number of species, the fact remains that considerable variations may be found even among closely related members.

Many birds conduct their courting in a simple and more or less inconspicuous manner; to any except an expert observer, their behavior during mating time may appear to differ very little from that normally displayed. A great many others, however, have developed distinctive—and often spectacular—eye- and ear-catching practices. These may range from nothing more than rather monotonously repetitive and not always musical calls, to some of the most elaborate, colorful, and complex displays, song-and-dance performances, and rituals found anywhere in the animal kingdom.

Because of its diversity, and the tendency of many species to combine vocal with visual means in wooing the female, avian courtship is difficult to categorize. There are, however, certain broad and general characteristics in much of the prenuptial behavior that make possible—not for any scientific purposes, but for convenient comparison—the grouping of species according to the means and methods employed by the males in winning their mates. Even though areas of behavior overlap in many instances, there are numerous birds that rely mainly on specialized traits or features. Thus the courtship of one group, such as that including our favorite backyard and garden songsters, may be almost entirely vocal, while in another, the pheasants and grouse, for example, the males depend primarily on the display of their resplendent ornamental plumes or on elaborate dances to impress the female. Males of some species court by building nests and offering them to the female as an enticement to mate. And then there are those whose courtship depends neither on beauty of voice or plumage nor on the offering of a nest; especially among birds whose sexes look alike, it often involves a ritualistic ceremony of movements and postures in which male and female are more or less equal partners.

For all those species whose males have, or grow at breeding

time, specialized plumes lacking in the female as well as in immature males, courtship usually is a one-sided affair conducted by the male bird only. In poses designed for maximum effect, such males display their often spectacular plumage to the typically drab-looking female. By the same token, breeding-time activities of these males are more or less restricted to courtship and mating; only a very few help with such chores as nest-building or the rearing of the young. Most of these males are highly polygamous and—with the exception of those that gather and defend against rivals a harem of females—have only a brief relationship with each of the mates they succeed in winning. The most promiscuous polygamists are invariably found among species with the most ornate and spectacular breeding plumage.

Male and female of the Red Jungle Fowl, a polygamous species

Male and female of the Canada Goose, a monogamous species

At the other end of the scale are the true partnerships, in which the mated couple faithfully share the responsibilities for the nest and the rearing and the protection of the young. Such bonds are almost always characteristic of species whose sexes look so much alike that it is often impossible to distinguish the male from the female on the basis of appearance. Most of these birds are at least seasonally monogamous, and some mate for life; among species noted for an especially strong and lasting pair bond is the handsome Canada Goose, probably the best-known, best-loved and most appealing of North American waterfowl.

A few species, such as the ploverlike phalaropes and the peculiar South American tinamous, primitive birds that look somewhat like quails, deviate from all behavior normally associated with avian courtship and breeding habits. Both the appearance and the roles of the sexes are reversed in these species, the female being not only the larger or more brightly colored of the two, but also the one that does most of the courting and leaves the task of incubating the eggs and raising the young entirely to the male.

Extensive research into the courtship practices of birds during the past few decades has yielded much information not only about such exotic and difficult-to-observe species as bowerbirds and birds of paradise; it also has revealed many new facets of courtship behavior as practiced by numerous well-known and familiar birds of woods and gardens, lakes, ponds, and seashores. Even though our knowledge in this field is therefore much more comprehensive now than it was forty or fifty years ago, there is still a great deal that remains to be learned about many rare, shy, or inaccessible species as well as about numerous common birds.

An increase in the sum total of our knowledge was only one result of research in recent years, which saw much of the emphasis in biology shifting away from the preoccupation with the purely functional, physiological aspects of living organisms, and toward a greater interest in behavior, including intangibles that cannot be measured by scientific tests or instruments. Newly discovered facts and insights gained through painstaking study and observation raised new questions while casting doubt on some firmly established theories, including the one that assumes animal behavior to be rigidly and inflexibly prescribed by inherited instincts. Apart from the fact that instinct remains a still-unexplained and mysterious force, it does not always seem to function as well and as widely as heretofore believed. On one hand, even humans, so lacking in many other instincts, seem to have better-developed inherited knowledge in at least one important area—that of unfailing recognition of its own species—than certain other animals including most birds. On the other hand, the supposedly largely human prerogative of learning to adjust and change behavior through experience appears to be more prevalent in animals below the mammalian level than formerly assumed. Studies have shown, for instance, that learning and experience, rather than instinctive, inherited "knowledge," may equip certain birds with such vital skills as capturing food, recognizing and avoiding enemies, and, last but not least, perfecting the musical abilities necessary for the male's successful courtship.

It has long been known that the musical offerings of many of our familiar and favorite songsters belonging to the Oscines—the suborder of perching birds that includes the typical "songbirds"—may display considerable regional as well as individual differences and variations. Excellence of performance was correctly attributed to the presence of good "teachers"—highly accomplished older males from which the younger ones could learn by listening—but the degree to which the basic, characteristic song of a species was an inherited, instinctive talent remained in dispute. Those biologists who believed that every important phase of animal behavior is rigidly defined and limited by inherited instincts naturally assumed that birds are born with the ability to perform their species' courtship song. Long and painstaking experiments conducted mainly by several European biologists have proven this assumption to be only partially correct. These biologists raised songbirds of various species from the egg to sexual maturity without once permitting them to hear the specific song of their own kind—or, for that matter, of any other kind. The birds so raised were grouped according to their performances: Those of the first group were indeed capable of performing the entire song of their species without ever once hearing it, thereby

The Lesser Whitethroat of Eurasia

These two small European song-birds typify the great difference in the degree of instinctive singing ability. The Whitethroat can give a note-perfect rendition of the song of its species without ever having heard it; the Chaffinch has to learn from older males.

The Chaffinch of Europe

proving that it was an entirely instinctive, inherited talent. Birds in the second group, however, could utter only a few rudimentary portions of their song, and were incapable of perfecting it until after being permitted to listen repeatedly to an accomplished older male of their species; individuals without such a chance remained very poor singers.

The results of these studies are fascinating because they furnish additional proof that the development of some abilities that play a vital role in a bird's life may be left at least partly to social contact and learning experiences rather than to rigidly fixed inherited knowledge. The considerable degree of variation in individual song performance, as well as the existence of a kind of "musical tradition," through which individual embellishments and variations of a melody may be passed on from one generation to another in some species, bears witness to such acquired musical skills. This greater freedom of expression apparently outweighs any possible disadvantages of lacking inherited ability.

Naturalists studying the field of sex recognition among animals have ample reason to believe that most mammals are equipped with an inherent ability enabling them to distinguish an individual of their own kind from any other when mating time approaches. This applies even in those instances where young mammals are raised without the company of others of their species, a fact incorporated—fancifully but probably accurately—by Rudyard Kipling in his *Jungle Book*. Mowgli considers himself a wolf—until love comes to him and forces him to leave his "brother" wolves and return to humankind. So far as we can tell, most birds could not have done this, for with notable exceptions—brood parasites such as cuckoos, for instance—they apparently lack that particular instinctive knowledge. In case after case, careful studies of birds raised without the company of others of their kind have shown such orphans to be totally confused about their true relationship later on in life, and even after sexual maturity incapable of distinguishing their own not only from other bird species but also from other animals belonging to entirely different groups. One of the most striking as well as pathetically comical instances of this was reported by Konrad Lorenz, noted German student of animal behavior. It concerned an albino peacock in an Austrian zoo, the lone survivor of a brood that had succumbed to a spell of bad weather. The peafowl chick was placed in the only warm room available during the meagerly funded postwar years. That room just happened to be the one in which the giant tortoises were housed. Although the young peacock flourished in these surroundings, the peculiar effect of its reptilian roommates on the bird became apparent not long after it had attained sexual maturity and grown its first train: beginning then and forever after, the peacock displayed his magnificent plumes in the amorous "wheel" position *only* to giant tortoises, eagerly if vainly courting these reptiles while ignoring even the most handsome peahens with which the zoo supplied him. Apparently, once a bird becomes "fixed" (or "imprinted," as biologists call it) on the wrong kind of mate, it is all but impossible to reverse this process. In the case of the albino peacock, lack of instinct condemned the unfortunate bird to a life of involuntary celibacy.

Lorenz tells of similar experiences with different birds he kept as pets over the years. Thus a goose—also the lone survivor of a brood—which had been raised with chickens formed a strong attachment to a rooster, who of course ignored her, while she in turn ignored the gander Lorenz had acquired especially for her.

The German naturalist's most extensive and detailed studies of such avian fixations, or imprints, involved the small Eurasian members of the crow family known as Jackdaws, which combine the intelligence of their larger relatives with a much more appealing and attractive personality. Gregarious birds, Jackdaws have a

highly developed social life, and become easily attached to human beings who raise them, as Lorenz found out when a young male Jackdaw approaching sexual maturity began to court him, trying to stuff love offerings of food into the naturalist's mouth. Lorenz was especially intrigued by the bird's ability to arrive at the correct conclusion that the human mouth is the equivalent of the avian beak, a feat that only served to highlight the complete lack of instinct displayed by the Jackdaw's impossible choice of a mate.

These and many similar observations seem to offer conclusive proof that most birds do not have the inherent ability to distinguish between individuals of their own and other animal species, and that they learn to do so only through social contact with their own kind at an early age. Self-evident exceptions to that rule are the brood parasites—cuckoos, cowbirds, and others—which invariably are raised by foster parents belonging to other species, and never see individuals of their own until they are full grown; they obviously must have the built-in ability of recognition that is missing, or at least deficient, in most other birds.

Since the typical young bird survives only if reared by its parent or parents, and therefore by necessity grows up in the company of others of its own kind, the lack of the "recognition instinct" is unimportant for all practical purposes. Its interest to biologists rests mainly in the fact that it illustrates the varying degrees of functioning instincts affecting different areas of behavior in a variety of species.

Another result of behavior research has been the reassessment of the part played by the sex drive, which formerly was believed to be all-encompassing—the very "life force" itself—by those biologists to whom every living organism was merely an aggregate of physiological functions and chemical reactions. Observations of numerous animals including birds have shown, however, that although the sex drive may increase, enhance, and heighten certain traits, talents, and abilities for use in courtship, it does not produce them. Thus the courtship songs of many oscine birds are often only portions of the complete—and frequently much more complex and varied—song characteristic of those species. This full "specific" song may be heard when sung, softly and for its own entertainment, by the young bird before it is sexually mature, and then again after the breeding season is over and the sex drive becomes dormant. Similarly, the peculiar, ritualized dancing performances and greeting ceremonies that characterize the courtship of various species are not always restricted to the mating season; they also may serve as a form of social contact and entertainment, or simply as an expression of exuberance, at other times of the year.

By and large, avian courtship practices seem to be just specialized, and often spectacular, parts of a greater and much more complex behavior pattern, much of which is still incompletely understood. A number of other drives, forces, and powers-behind-the-scenes, such as that producing the pronounced social caste or rank system known as the pecking order, play parts no less important than the sex urge. A case in point is the Passenger Pigeon; it is believed today that the indiscriminate slaughter suffered by these doves in the past century was only indirectly responsible for their becoming extinct. Theoretically, at least, there were more than enough birds left to reproduce and preserve the species. It appears, however, that for some reason these pigeons could breed successfully only in huge nesting colonies, and that once these colonies were broken up, the smaller groups, although sexually mature, fertile, and healthy, were no longer capable of propagating their kind owing to some still unknown factor or factors.

Among the outside influences that guide and govern the lives of most birds, none is more important than light. As can be seen from the size of a bird's eye and visual center in the brain, the sense of sight is most highly developed, closely followed by that of hearing, whereas the sense of smell is generally poor. Most mammals share with birds only the acute sense of hearing, but differ sharply in having a superior sense of smell but poor vision, frequently being all but colorblind.

Seasonal changes in light govern much of the average bird's life, telling migrating species when to start their flights, activating the production of sex hormones, and providing a clock for all species with diurnal habits. Studies and experiments with songbirds have shown that the increase in the length of the day stimulates the pituitary gland, which in turn triggers sex hormone production by the gonads. These sex hormones affect the song centers in the male's brain, resulting in the beginning of the courtship songs. It is, of course, widely known that artificial light can cause diurnal birds to sing at night, male songbirds to voice their courtship melodies in winter, and hens to lay eggs regardless of the season.

It would be hasty, however, to conclude from the preceding that light is the vital key to the courtship activities of *all* birds; one only has to think of the many nocturnal species to realize that light cannot possibly be a factor in their breeding cycle, since they court and mate in darkness, during the night. The many birds living in equatorial regions where there are no seasonal changes in the length of the days also have to rely on other factors for stimulation of the sex hormones. It has been observed that a "population explosion" among certain insects such as locusts triggers the beginning of courtship performances of numerous tropical birds, including some of the brilliantly colored, iridescent African starlings, that rely heavily upon these insects for food. And then there are a few species, among them the lyrebirds of Australia and the Emperor Penguin of Antarctica, whose breeding cycle is set in motion by the *decrease* in the length of the days.

Since it is normal and typical for all animals to utilize their natural traits and attributes in courtship, it is not surprising that birds, probably the most universally vocal of all major animal groups, should employ their voices extensively during the process of wooing their mates. Few birds, even those that cannot sing in the conventional sense, and whose voices may sound anything but pleasant to the human ear, are completely silent during courtship. This general rule is confirmed by the usual exceptions, for there are a few birds that let their voices be heard at other times, but never during courtship. Generally, however, the breeding season is the time when not only the "voice of the turtle(dove) is heard in the land," but the voices, melodious or unmelodious, of most other birds as well. A great number of birds, especially among the oscine or songbird group, court predominantly or even entirely by song, which apparently for them is an effective way not only of attracting a mate but also of claiming a territory, warning off potential rivals, and later on encouraging the female while she broods the eggs. It should be noted in passing that many oscine birds are actually very poor singers, whereas other species not belonging to the "songbird" group may produce very pleasant melodies.

The families of thrushes, finches, larks, and sparrows include the most accomplished of all songsters. Many of the finest among them, such as the renowned European Nightingale, the North American Mockingbird, and the Wood, Song, and Hermit Thrushes, are distinguished by the similarity of the sexes in both size and coloration. Even where sexual dimorphism—a pronounced difference in the appearance of the sexes—exists within these groups to the extent that the male wears the brighter colors, it seems to play only a secondary role in courtship behavior.

The great majority of songbirds, such as the American Robin pictured earlier, perform their vocal artistry while perching, often selecting a spot from which they can survey their domain. Most people are familiar, not only with robins, but also with starlings or

song sparrows sitting atop a telephone pole, an antenna, or some other such elevated spot, feathers fluffed, wings drooping, caroling their song of love and challenge. Other songbirds prefer to remain hidden—it is not easy to observe a nightingale during its masterful performance, despite the fact that, its name notwithstanding, this bird sings as freely during daytime as it does during the night. Nightingales usually sit quietly in a bush or hedge while pouring forth their mellifluous yet vigorous song, and their inconspicuous brownish-gray coloring makes them difficult to distinguish from their surroundings. There are a number of songbirds, however, that become articulate only on the wing; the most famous of these is the Skylark, which has inspired poets for centuries.

Birds not belonging in the oscine group but also depending mainly or entirely upon their voices for wooing the female include many species that can bring forth little more than simple if distinctive calls. The most widely known of these is the European Cuckoo, whose loud but monotonous two-syllable call, repeated over and over again, heralds spring in large parts of Eurasia. As far as can be ascertained, this call *is* the cuckoo's courtship, and apparently suffices to attract the female.

Among the birds whose courtship consists mainly of simple although often quite musical calls are the peculiar bellbirds of South and Central America. As the name indicates, their calls have a ringing, resonant quality that may carry over a considerable distance. Three of the four species are further distinguished by the

peculiar, black or gray, spike-shaped wattles growing from the base of the bill. The male of the White Bellbird attracts the female by its ringing calls, then displays with gaping mouth while sitting on the branch which represents its entire territory; the pure white plumage and the single, long blackish wattle growing from the base of its upper mandible distinguish this particular species.

Other birds court by uttering a variety of often cacophonous crying, wailing, or braying sounds few humans would consider pleasant, although the mournful wail of the Common Loon, which sounds somewhat like the howling of a wolf, does add a very special flavor to the nighttime atmosphere of the northern wilderness areas where these birds breed. The loud and crazy-sounding "laughter" of the loon, however, which also is a part of the courtship concert, affects most people very disagreeably; the origin of the popular expression "crazy as a loon" can probably be traced to that sound.

Many wading birds of the plover group, on the other hand, produce beautiful whistling or trilling sounds as part of their courtship performances; some of them are so melodious that they could well be described as songs. They are often uttered on the wing during courtship flights requiring complicated aerial maneuvers. Snipes and the closely related woodcocks are, of course, famed for such flights, the chief difference between the two being the melodious little song woodcocks trill while on the wing, whereas the peculiar sound that accompanies the courtship flight of snipes is produced by the air that is forced through the stiff tail feathers.

Courting on the wing is typical for entire groups of birds, especially birds of prey. Many hawks and eagles, for example, go through their courtship performances while circling high above their hunting territory. As soon as the male has succeeded in attracting a female, she partakes in these aerial shows; the cartwheeling mating flight of a pair of eagles is an impressive spectacle.

Aerial courtship among numerous other species is an all-male performance, designed primarily to show off the suitors' special colors or plumes, or both, to the perching female. This is true especially of the hummingbirds, whose males are typically endowed with erectile crests, collars, or earfans whose glittering rainbow hues span the entire range of the spectrum. Displaying these colors quite literally from the best angle is essential because they are structural rather than pigmental, and may disappear altogether when seen from certain angles. Such iridescence is the result of a complicated optical principle, by which submicroscopic layers of thin films spaced at certain distances in the feather tissues absorb some of the components of white light but selectively refract others, thereby creating incredibly pure and brilliant iridescent colors, which are, however, visible only from certain angles. It must be assumed that male hummingbirds are instinctively aware of this fact, and so arrange their fantastic U-shaped courtship flights that these gleaming colors are presented at the optimum angle to the perching female. Thus the male Frilled Coquette of Brazil manages to show off a red crest and an iridescent green throat, as well as striking, white, green-tipped fans extending on both sides of the neck, to the demure-looking female during his courtship antics.

Other birds display not so much their colors as their specialized plumes during such flights. This is true of certain weaver finches whose males bedazzle the female with erratic aerial acrobatics while trailing tail plumes several times the length of their bodies, a practice the female evidently finds irresistible. Similarly, the Pennant-winged Nightjar of Africa courts on the wing; the male's last molt before the breeding season produces two-foot-long outer primaries, which look most peculiar on this barely foot-long bird. That these outsized wing feathers are in fact a nuisance seems to be borne out by the observation that they are broken or bitten off immediately after mating, presumably by the bird itself.

For the Pennant-wing, one of the rare birds that appear to be voiceless, the courtship flights take the place of the loud and persistent courtship calls typical for the rest of the goatsucker group. These well-known "chuck-will" or "poor-will" calls may be repeated so often during the course of a single night that they can drive human listeners to distraction; naturalist John Burroughs counted one bird calling 1,088 times in succession! With the exception of the Common Nighthawk of North America, which calls on the wing and underlines its courtship call with a flight performance, all other nightjars call from their typical sitting or rather crouching position lengthwise on a horizontal tree limb or similar perch. Also noted for their nighttime courting by distinctive calls are the predominantly nocturnal owls, which are structurally more closely allied to the goatsuckers than to other birds of prey.

Relatively few of the courtship practices mentioned in the preceding pages can be described as spectacular from a human point of view. Even those that deserve such a designation—the eagles', snipes', and hummingbirds' flight performances—are either too difficult to observe or too fast-moving for the human eye to appreciate their intricacies. Only slow-motion films, which of course distort the movements of these displays, can do some justice to the individual details of such aerobatics. The most impressive instances of courtship behavior, some of which are described and pictured in the chapters that follow, are generally those that afford the human observer a chance to distinguish and enjoy each phase of the various ways in which birds go about their courting. In observing the display artists, the song-and-dance experts, the working suitors, and those performances that feature male and female as equal partners in often complex courtship rites and ceremonies, it becomes clear that these prenuptial activities are part of the individual species' self-expression that goes far beyond the immediate, necessary goal of attracting and enticing the opposite sex.

2. *Plumage Pageantry*

GLEAMING GOLDEN in the light that filtered through the canopy of the jungle trees, the long, filmy flank plumes of the Greater Bird of Paradise cascaded over its back as the bird danced and pirouetted on the sloping display bough. Each movement was timed exactly to coincide with those of another male performing on a different branch of the same tree. As the two birds moved, they called to each other in resonant, bugling tones at intervals spaced increasingly closer together as the calls were synchronized with the stepped-up pace of the dance, which became even more frantic when a female joined the crowd of observers consisting mainly of younger males with incompletely developed plumes, most of which made some excited but futile attempts to emulate the displays of the two star performers. Finally, the speed of the movements abated as the dancers slowed down and began to shuffle along their respective branches, dragging their legs as though crippled, with lowered heads and drooping wings, the flank plumes still cascading over their backs in filmy clouds of yellow. The climax of the dance followed shortly thereafter; crouching low now on their boughs, legs tucked out of sight beneath their chests, wings drooping, the males seemed to go into a trance. With their glassy, staring eyes and gaping mouths, their bodies trembling with spasmodic shivers, they appeared close to expiring even as they raised that gorgeous fountain of cascading plumes to its utmost height, and were thereby transformed into exotic blossomlike objects that had lost all resemblance to birds. As the female and the other birds watched, both performing males maintained that incredible pose for the better part of a minute. Only then did they relax and let the plumes fall back into their normal position; leaving their display boughs, the two exhausted dancers settled, each on a separate branch, to recover from their exertions. At this point, there was no attempt to pair off with the female for the purpose of mating, for theirs had been only the first in a sequence of displays that are sometimes repeated several times in a row.

Relatively few persons have had the chance of observing the spectacular dances of New Guinea's birds of paradise. Efforts to keep and raise them in captivity have been generally unsuccessful, and many of the most striking species are rapidly disappearing because their original habitats are being destroyed as man constantly expands his needs for living space. Others are so rare that they remain largely unknown to this day, and have never been captured alive or even observed in the wild; a few mounted specimens to date are our only source of information for some of these rare species.

The extreme dearth of reliable knowledge, until a few decades ago, about the habits of these birds was not due to ignorance of their existence; they were well known to the natives of New Guinea and the surrounding islands, and hunted for their marvelous plumage, long before Europeans ever came to those parts. When the *Victoria*, the ship of navigator Fernando Magellan, returned to Spain in 1522 with some prepared skins as a native ruler's gift to the king, the Spaniards could hardly believe their eyes, so otherworldly did the exquisite creatures seem to them. Even the scientists who classified the birds later echoed that sentiment by naming the group Paradisaeidae. In assigning heavenly origin to them, Europeans, perhaps unwittingly, followed the Malay example, for Portuguese traders had come to know the birds of paradise as *manucodiata*, which is a corruption of the Malay words *manuq dewata*, meaning "birds of the gods."

One of the most fascinating aspects of this group of forty-odd basically crowlike birds is the wealth of diversity in feather modification. Practically every group of feathers found on a bird's body has been modified into ornamental plumes in at least one species of the Paradisaeidae. This applies to the tail and wing feathers as well as to those of the head, nape, back, throat, flank, and breast. In addition, the types of modification show the same wealth of variation. The ornamental plumes may be filmy and lacelike, velvety, or endowed with metallic iridescence. Some are drawn out into thin wiry threads; others resemble tiny flags or vanes that have a hard, enameled appearance.

The relatively few skins that reached Europe in the seventeenth century gave rise to the legend that these birds had no legs, nor did they need them, since they allegedly were always on the wing, and even the eggs were incubated by the female on the flying male's back. This fanciful tale originated through the native taxidermists' habit of removing the birds' legs when preparing them; it was immortalized by the Swedish naturalist Karl von Linné—the so-called father of modern biological classification—when he named the Greater Bird of Paradise *Paradisea apoda*, the "footless paradisean."

The exact geographical source of the skins that aroused so much excitement and admiration in Europe was not known for certain until the early years of the nineteenth century, and reliable scientific studies were not begun until several naturalists, most prominent among them the Englishman Alfred Russell Wallace, described several species and their habitats just about a hundred years ago. Wallace also brought back from his travels specimens of some newly discovered members of the group, including *Semioptera wallaceii*, the Standard-winged Bird of Paradise, whose scientific name honors its discoverer.

Toward the end of the past century, the demand for fancy plumes for women's hats and dresses was in full swing; this modish trend proved disastrous to a great many birds with ornamental plumes including the beautiful "birds of the gods." As many as 50,000 skins per year are reported to have been shipped out of the New Guinea archipelago in the 1880s and 1890s. By the time this fashion-inspired madness had abated twenty-five years later, inestimable damage had been done, and many formerly common species from various parts of the world had become rare.

Although Wallace contributed much sound information to the then very small store of reliable knowledge about these birds' habits, it cannot be established whether all his descriptions stem from firsthand observations; much of what he reported, however, has been confirmed by recent research. Because large parts of the dense jungle environment in which the birds of paradise make their home and conduct their fascinating displays are all but inaccessible, only a handful of zoologists have been fortunate enough to observe these birds in the wild, and even then only after braving long, uncomfortable and mosquito-ridden stretches in the confining quarters of blinds erected in the jungle. From these observations, however, it has been possible to piece together the story, both in words and in pictures, of the incredible display dances performed by the males during the breeding season, which must rank among the most spectacular as well as the least understood of all avian courtship rituals.

Prominent among ornithologists who devoted much time and effort to the study of birds of paradise was the late Thomas E. Gilliard, a former Curator of Birds at the American Museum of Natural History in New York. In 1964, he headed an expedition whose goal was the study of the Greater Bird of Paradise, especially its courtship performances. To achieve this goal, the American biologists did not, however, have to travel halfway around the world to the Aru Islands of New Guinea; instead, they headed for Little Tobago Island in the Caribbean off the northern coast of South America. Into this equatorial region, whose climate and steaming jungles are very similar to those of their ancestral home, the Greater Bird of Paradise had been introduced by English bird lover Sir William Ingram in 1909. Conditions on the island apparently suited the birds for over the decades they flourished.

After long, hot, and frustrating waiting periods in their blinds in the jungle, Gilliard and his associates were finally rewarded by the spectacle of magnificent performances by a number of adult males, usually played to an audience that included at least one or more females as well as a number of young and incompletely plumed males, which latter avidly followed the performance of the star actors, and excitedly tried to emulate them while remaining at the periphery of the main stage.

What the biologists were able to capture on film confirmed earlier and often exaggerated-sounding reports. Here indeed were the "entire trees filled with waving plumes" (described by Wallace a century ago), first-hand observation of the fact that these birds engage in group courtship performances whose chief prerequisite seems to be not so much the presence of females but rather that of at least several other males. Why this should be so is as incompletely understood as the entire complex, bizarre, and, for the performers, exhausting ritual.

The main mystery of these sophisticated prenuptial dances centers on the fact that they often take place in the absence of the females, are triggered by the gathering of two or more fully plumed males ready for display, and are stimulated by an audience consisting largely or entirely of other males. The presence of females, or their appearance during the performance, does have an impact on the displaying males, however, for it results not only in more frenzied dancing but also in prolongation of those periods during which the males maintain the rigid, trancelike poses that distinguish the climax of the performance, as described earlier. Although these flowerlike display stances are believed to be a form of symbolic mating, actual mating does not seem to follow quickly, at least not always. Gilliard observed some males going through the entire sequence four times in succession without attempting to leave the scene of the dance and join a female. Mating evidently does not take place in or near the display tree; exactly where, when, and how long after the courtship rituals these birds made has still not been established with certainty. It is believed, however, that the Greater Birds of Paradise, as well as most of the other unusually ornate members of this group, are polygamous in a very special, promiscuous way. The male apparently meets each female only fleetingly for the sole purpose of mating, and then goes on to another performance and, eventually, another mate. This behavior is, of course, typical of numerous species whose ornately plumed males present a striking contrast to the ordinary-looking females, but differs considerably from the kind of polygamy practiced by the males of such species as grouse or pheasants, which may gather, lead—and guard during the breeding season—harems made up of several females.

It seems clear from all studies and observations of the Greater Bird of Paradise in its natural habitat that the spectacular performances of this and closely related species go far beyond a courtship ritual designed merely to attract and win the female. Social contact and competition with other males, the stimulating and exciting effect of such contests, and the presence of more or less participating audiences are probably just as important a part as the wooing and winning of a mate. Despite the filmy, lacelike quality of their ornamental plumes, these birds are really quite virile and masculine; their bugling calls are strong and resonant, and their dancing frequently borders on the violent. Gilliard observed repeatedly that those males with the longest and most beautiful plumes could and did claim the center of the stage; birds with less perfect plumage regularly and without a fight yielded their places to the more resplendent individuals even when the latter arrived late during the display. These observations lend strength to the theory that the dance performances may be a substitute, at least to a certain extent, for the fights that males of many other species such as pheasants and grouse have to go through in order to establish a "pecking order" or caste system, in which the superior males may claim the biggest share of the available females.

Despite such partial explanations of the rituals, no one claims they answer all the questions about the courtship habits of these fabulous birds. This is all the more true because not all of them by any means indulge in the social dancing typical of the Greater Bird of Paradise; even some closely related species within the genus *Paradisea* are more or less solitary and content with performing for the female only. They also each have their own variations of the display performance; thus the Lesser Bird of Paradise slowly tilts forward toward the end of its display until it hangs upside down from the branch, and the exquisite Blue Bird of Paradise begins its ritual by suspending itself in this manner. This species is the only one of the genus *Paradisea* in which blue predominates in the plumage, and a marvelous spectacle it presents surrounded by a filmy blue cloud of cascading flank plumes as it hangs from its perch.

Although most members of the Paradisaeidae go through their courtship rituals above ground, and often high up in trees, this characteristic habit has its notable exceptions. The Magnificent Bird of Paradise, a small species only eight inches long, descends to the ground and clears a "dance stage" that may measure fifteen feet in diameter. It then snips off the leaves of any sapling growing within this circle, and performs its display acrobatics for the watching female while dancing up and down the sapling in an almost vertical position. The gleaming iridescent green chest feathers are puffed out so they can catch the unobstructed light, and a brilliant fan of yellow neck feathers is expanded to its maximum width.

The Six-wired Bird of Paradise—so named because of the six wirelike plumes topped by tiny black rackets that grow from its head—does not bother with trees or branches at all. Like the magnificent, it clears a dancing area on the ground, removing debris and snipping off any small plants that may be in the way, until a roughly circular "stage" is free of obstacles. The bird then performs its dance on the ground for the female watching from a nearby branch. During the ritual, the male spreads his elongated chest feathers until they form a kind of ballet skirt accentuated by a patch of brilliantly metallic feathers above. The total effect is complemented by the headdress of waving, nodding, flaglike plumes.

Some ornithologists believe that those members of the Paradisaeidae that perform their courtship rituals on the ground rather than in trees represent the group's evolutionary trend toward ground display; this theory is probably based at least partly upon observation of the closely related bowerbirds of New Guinea and Australia, whose exceptional courtship practices will be pictured in a later chapter.

In addition to the birds mentioned above, the Australian region hosts so many unusual, unique, and marvelous animals that it does not come as too much of a surprise to find in Australia a family of birds consisting of only two species with no close relatives anywhere else in the world. Australians are justifiably proud of their two lyrebirds, and picture them on many stamps and seals. Although not at all brightly colored, the males in courtship display are striking indeed.

Lyrebirds possess so many anatomical peculiarities that they have caused considerable trouble to ornithologists trying to find the proper niche for them in the biological system of classification. At first, the brownish, rooster-sized birds with the long tails of modified plumes were believed to be pheasants; upon closer study, however, their voice box was found to resemble more closely that of the oscine, or songbird, group, even though their weak breastbone is more like that of water birds. Finally, although classified as oscine birds, the two species, known respectively as the Superb and Albert's Lyrebird, were considered different enough to be placed not only in their own separate family but also in a suborder shared by just two other species of rare Australian birds.

The best-known of these avian mavericks is the Superb Lyrebird, sometimes also called the Common Lyrebird. It is familiar to many persons from pictures in which the male is shown with its tail held erect and forming the famous "lyre" which gave the two species their name. Picturing this pose as typical, however, conveys a somewhat misleading impression, because it is never maintained for more than a few seconds when it occurs fleetingly during courtship display.

The only striking feature of these birds, whose coloring is predominantly brown and gray, is the tail of the male. It not only is very long, far exceeding the body length, but also consists of highly modified plumes that have lost all resemblance to ordinary tail feathers, and with it the capability to function as a rudder during flight. This latter fact, however, is hardly a drawback for birds that are largely terrestrial and much prefer running and leaping to flying anyway.

When fully erect, the Superb Lyrebird's outer pair of tail feathers form the frame of the "lyre"; these feathers are gracefully curved, banded with dark brown, and tipped with black. The next six pairs of tail feathers are totally modified; through the lack of barbules (the hooked processes of the typical feather branch that unite the individual branches into a firmly meshed web) they form loose, filmy plumes that are brown above, silvery white below. The two innermost feathers are very narrow, ribbonlike, and light-colored. Although the tail of the somewhat more reddish Albert's

Lyrebird is similar to that of its better-known cousin, it lacks the distinctive feathers forming the frame of the lyre.

Except during courtship, male lyrebirds carry their tails folded horizontally very much like pheasants; only in this fashion can they make their way through the dense underbrush of their homeland without damaging the delicate plumes.

Lyrebirds establish large territories and tolerate no other males within the boundaries of their domains; the resident females are claimed by the dominant male. Despite the birds' extreme shyness and caution, a few patient observers have succeeded in witnessing the courtship performance of both species in their native surroundings. The Superb Lyrebird sets the stage for its prenuptial display by raking leaves and soil into a mound about three feet across; Albert's Lyrebird scratches out a shallow crater of approximately the same diameter. A strong, virile male may establish as many as a dozen separate display grounds within its territory.

The courtship ritual of lyrebirds begins when the male climbs atop his mound—or enters his crater—and proceeds to sing from his extensive repertoire. After a minute or two of this purely vocal entertainment, the bird unfolds its tail, and slowly raises it over the back. At that point, the tail of the Superb Lyrebird fleetingly assumes the position that has given the birds their name, with the

curved outer pair of tail feathers forming the frame, and the filmy inner plumes the strings, of the musical instrument. The male then continues to tilt the tail forward until the loose, filmy plumes, whose undersides gleam silvery in the light, cascade to the ground over its lowered head, completely obscuring the bird's body beneath. Despite the generally muted colors of the lyrebird's feather dress, there are few displays in birddom more impressive than this peculiar performance.

As mentioned earlier, it is not easy to observe these birds during their courtship even though their voices may be heard tantalizingly nearby. Because of their special vocal artistry, getting to hear them is no problem at all, for one of the outstanding features of these wonderful birds is their talent for vocal mimicry, said by some who heard them to surpass all other famous avian mimics including the renowned mockingbirds.

After the climax, the lyrebird utters a few sharp notes, folds its tail, and stalks away to the next mound and the next performance;

Male Superb Lyrebird with its tail folded in the normal position

if not immediately, he can expect to attract and win several females in the course of his rounds. After mating, the females are on their own; the males leave the task of both nest-building and rearing of the young entirely to their mates, each of whom spends an entire month constructing a large, bulky, domed nest that has a side entrance. In this oversized nursery, the female incubates her single grayish egg.

Rivaling or even surpassing the lyrebirds in shyness is the Great Argus Pheasant, another spectacular performer who makes his home in the jungles of the Malayan Peninsula, Sumatra, and Borneo. This pheasant is a relatively little-known member of one of the most familiar and widely domesticated group of birds, which includes the ancestors of our barnyard chickens. Pheasants are typically distinguished by the elongated decorative feathers, especially of the tail region, that adorn the males, and their bright and often beautifully metallic, iridescent color patterns. The Great Argus Pheasant, one of the largest members of the family, and a bird described by Thomas E. Gilliard as "almost too beautiful for

Male Great Argus Pheasant

words," deviates from the majority both in the type of feather modification and in their coloring. Until recently, very little reliable knowledge existed about the habits of this magnificent bird—which may measure up to seven feet in total length—although it was better known than its relative, the Crested Argus Pheasant. The latter is distinguished by having the largest feathers of any wild bird, the central pair of tail feathers being six feet long and six inches wide. This fantastic bird is believed by many scholars to be the prototype of the Chinese Feng-Huang, later known as the Phoenix in Egyptian lore. Other experts, however, opt for such birds as the Golden Pheasant, the Great Argus Pheasant, and the Common Peacock as being the sources of inspiration for the fabled bird that rises rejuvenated from the fire into which it plunges itself when it grows old.

Even without the nimbus of fables and myths, the Great Argus Pheasant is unusual enough to warrant special attention and inter-

est. As a matter of fact, the bird, or rather the pattern of the feathers displayed prominently during courtship, had a major role in one phase of the arguments about Charles Darwin's evolutionary theory. The core of the controversy, which centered on the artistic perfection of that pattern, becomes readily understandable to anyone seeing the bird both in its normal and in its display position.

As mentioned earlier, the Great Argus is spectacular for the size, type, and pattern rather than for the coloring of its ornamental feathers. Delicate browns and grays predominate in the plumage; the female is an inconspicuous brownish bird that would hardly rate a second look. The male, however, is impressive even in its normal, nondisplay position. The center pair of tail feathers measures almost twice its body length, and the greatly elongated and broadened secondary flight feathers—a rare type of feather modification—have an intriguing pattern. Finely penciled with black on a brown background, the feathers display a chain of highlighted eyespots, graded in size and running down the center along the shaft. These eyespots, which are in the center of attraction during display, have puzzled naturalists ever since they attempted to assign specific evolutionary explanations to the various physical features and characteristics of animals.

The most extraordinary aspect of the eyespots is their almost uncanny three-dimensional effect, for they are highlighted and countershaded in a way that makes them appear as protruding, globular objects—but only if the light is coming from the right angle. Darwin, who had heard much about this startling plasticity of the eyespots before he ever saw the feathers, was at first disappointed at how flat they looked to him. The friend who was showing him the prepared skin quickly corrected that impression by arranging the feathers in the position in which the light would normally strike them during the bird's courtship display. Darwin immediately saw then what others had seen before him: the optical illusion that turns the eyespots into three-dimensional, spherical objects, an illusion so strong that observers feel tempted to touch the feathers in order to make sure that they are indeed flat surfaces.

Providing a plausible evolutionary explanation for the Great Argus Pheasant's three-dimensional feather markings presented a sticky problem for Darwin and his followers. The optical perfection of the eyespots involves an incredibly complex, delicately engineered and coordinated arrangement of different pigments in the thousands of tiny individual branches that make up each of the feathers; it was hardly possible to assign so sophisticated a pattern to the catchall category of evolutionary "accidents" or by-products. Darwin conceded this, but sought to advance the theory that the eyespots must have evolved slowly through selection—by discerning Argus females—of males with the more perfect and impressive "eye" markings. Apart from the fact that it leaves wide open the question of the pattern's origin, this theory presupposes the Argus female's ability to make rather sophisticated artistic judgments, a premise that other naturalists found unacceptable. Darwin's idea therefore has long been abandoned, and so has any real attempt to explain the origin of the pattern—or, for that matter, of other distinctive animal patterns, since nothing in the evolutionary theory of life provides a basis for explaining how and in what way such patterns originated. Instead, newer efforts concentrate on interpreting the *meaning* of the patterns and their role in the animal's life.

For the start of the courtship performance, whose climax presents the now-famous optical illusion, the male selects a clearing in the jungle, removing all debris and snipping off obstructing plants in the usual way of birds setting the stage for their prenuptial dances. The actual display is preceded by his uttering a series of rather harsh and unmusical sounds; then he begins to spread his impressive wings and cavort before the female. At the climax of his performance, the Great Argus slowly raises his wings and brings them together over his head until they form a complete circle of feathers behind which the bird's body is all but hidden, leaving only a small hole in the center through which the male can see and watch the female. At the same time, he also erects the tail, so that the five-foot-long pair of central tail feathers becomes visible behind and above the longest pair of wing feathers. In this pose, the male, like the Greater Bird of Paradise and the lyrebirds described earlier, loses all resemblance to a bird; with its body and legs disappearing behind the screen of ornamental feathers, it is turned into a complex if beautiful optical pattern whose meaning is far from being clearly understood. Finally, after holding this climactic pose for a little while, the bird relaxes, folds its wings in their normal horizontal position, and once again is just a large pheasant with outsized feathers. The impressive display has to be repeated several times, however, before the female shows any signs of entering into a receptive mood.

In line with one more recent theory, the "three-dimensional" eyespots of the Great Argus are seen by one scientist as possibly symbolizing kernels of grain, which other males of the chicken and pheasant family are known to offer the female ritually during courtship. This theory considers the eyespots as one small but probably important part in a complex and as yet poorly understood behavior pattern, in which the male's display is at once proof of his vital energy—emphasized by his size, voice, feather pattern, and the vigor of his movements during display—and symbolic of a parental role, expressed in the offering of equally symbolic food through which the cock asserts his superiority and dominance over the smaller, and therefore childlike, hen pheasant. The largely optical stimulations of the display, as well as the secondary acoustic stimulations of the sounds uttered by the male during his courtship performance, both usually repeated several times in succession, are seen in this view as affecting hormone secretion in the female, and thereby preparing the way for egg production and the readiness to mate.

A similar "synchronization" of the breeding roles of both sexes is believed to be the main object of all the spectacular plumage dis-

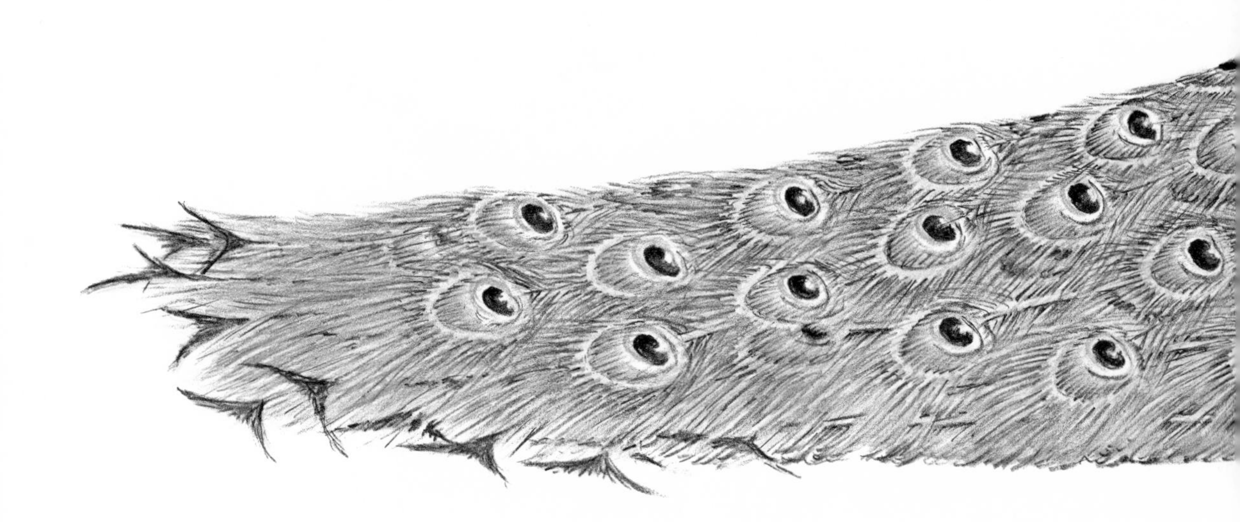

plays; the fact that definite esthetic values are involved makes the theory all the more fascinating, particularly in view of the human inclination to describe as an "animal" any person behaving in a coarse or brutish manner, especially in sexual matters.

One of the most beautiful of all avian displays, and one that has been studied more extensively than others because of the availability of domesticated stock derived from the wild birds of India and Ceylon, is the glittering show put on by the Common Peacock

Male of the Common Peafowl

during his courtship. In this display, the magnificent iridescent blue and green colors enhance the pattern of the male's highly modified train plumes, thereby turning his courtship ritual into a regal performance.

Although peacocks in their display poses are a familiar sight that can be viewed in any zoological garden, not very many people have followed the entire display sequence while at the same time observing the actions of the peahen—whom, after all, the spectacle is primarily meant to impress. During his performance, the cock executes a series of steps designed not only to present the fully extended fan to the female, but to do it in such a way that it will evoke the right response in his mate-to-be. Moving around so that his back is turned toward the hen, who at this point appears completely uninterested, he waits until she comes close to the edge of the fan. With a sudden, jerking movement, he then swings around

to face her, and tilts the now almost concave fan toward her with a seductive rustling of feathers and a multihued glitter of color. Regardless of how often one may have witnessed this display, only a totally blasé person could fail to be impressed anew each time by the long plumes spread to form their jeweled tapestry of burnished green dotted with those gleaming eyespots. The hen's rather peculiar response, however, consists of pecking at the ground as if in search of food. Even though to human observers this action seems to be a sign of indifference, and in no way connected with the courtship proceedings, it has an immediate effect upon the male, who is thereby stimulated to an increased effort. This seems significant in view of the fact that food and more or less symbolic feeding are an intricate part of many avian courtship ceremonies. The peacock's entire display sequence also may have to be repeated several times, however, before the hen is ready and willing to succumb to her suitor's charms and his regal wooing; here, too, repetition apparently paves the way for the female's receptive mood.

It appears that in at least some of the more spectacular instances of avian courtship, human observers are more readily impressed by the male bird's display than the females of its own species. During their studies of the peacock's performance, ornithologists often expressed a vague irritation as they recorded the seeming indifference of the peahen while faced with so much splendor. Such reactions, by experts, serve to highlight the difficulties of any attempt to "decode" the language of these prenuptial plumage displays; the more carefully they have been studied by naturalists, the more intriguing and often puzzling they have turned out to be. Once explained as being "simply" one of various methods of attracting the female, they are now recognized as highly sophisticated, complex behavior patterns. The coordination of movement, sound, and color effects necessary for maximum impact, the esthetic qualities involved, the birds' instinctive knowledge of how to present the ornamental plumes quite literally "from the best angle"—all these facts defy simple explanations. Most difficult to interpret and understand, however, are the many intangibles that work together to produce the synchronization of the male's dominant role with the more passive part of the female in order to achieve a mood conducive to mating and procreation. Since there is never any question of force, and the choice of the right moment is left entirely to the female as her prerogative, the elusive, intangible elements of courtship can only add to the fascination of the ornate displays staged by these wonderfully plumed birds.

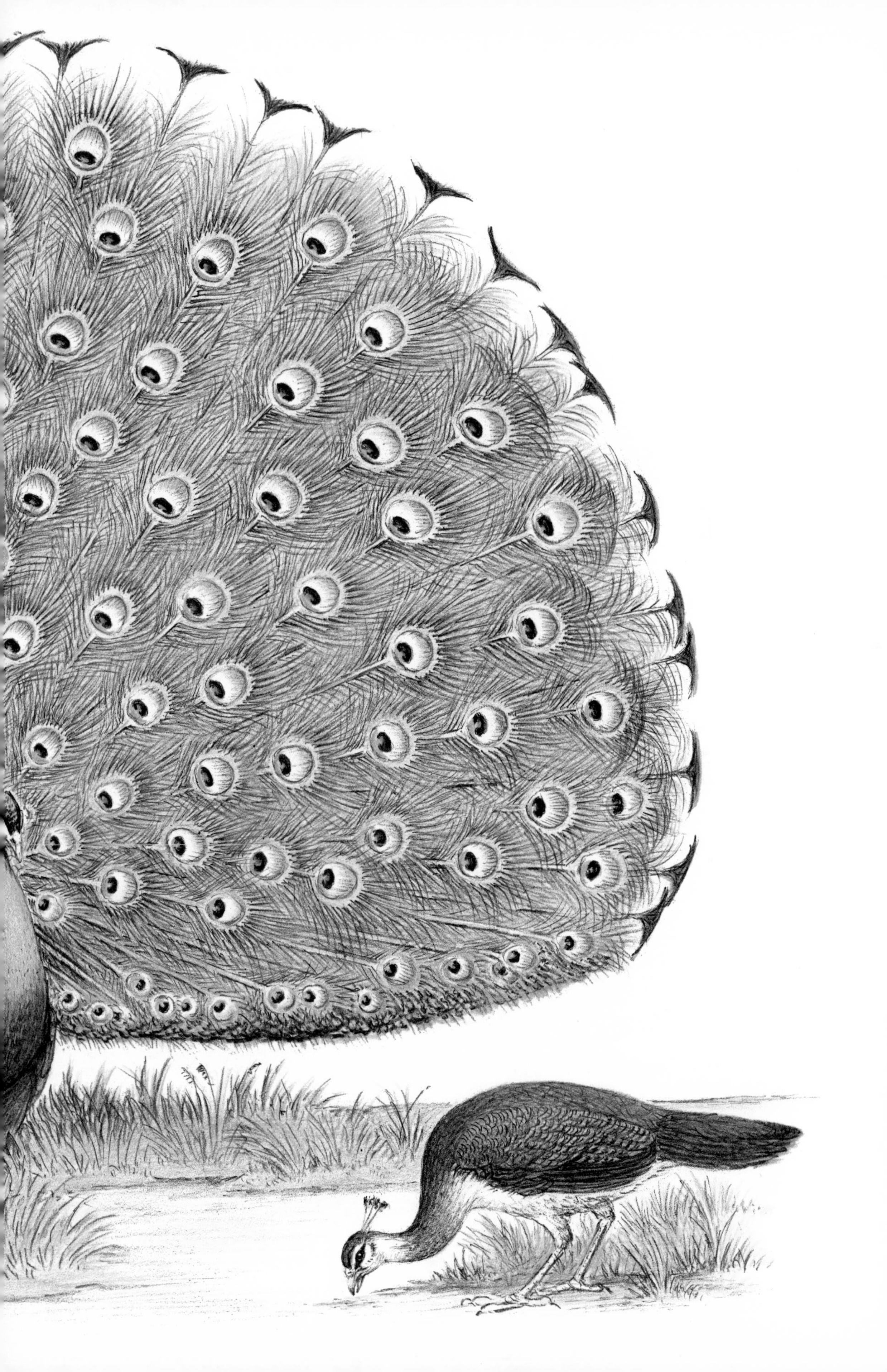

3. *Display Acrobatics*

THE PREDAWN QUIET of the gray, misty spring morning high up in the Swiss Alps ended abruptly half an hour before sunrise as a Black Grouse cock announced the beginning of his courtship performance. Soon other cocks joined him from the slopes and across the ridges, until the air was filled with the squeaking, hissing, rolling sounds that are the black grouse's song of love. As the sun rose above the horizon, its rays penetrated into the clearing chosen by the old cock who started off the chorus, and drew glints of steel blue from the black feathers of the now rapidly moving male, who bowed and scraped, danced, pranced, and pivoted with surprising agility for so heavy-bodied a bird. Finally, crouching low, neck extended and the head thrust forward, wings half spread and drooping, he lifted and extended the tail to form an almost lyre-shaped fan. The gleaming blue-black of the body plumage contrasted with the white bands on the dark brown wings and the white undercoverts of the tail, as well as with the fiery red wattles over the eyes. Throughout the dance, the cock uttered the distinctive sounds that proclaim to the surrounding countryside that he is in the mood for wooing.

In the meantime two other cocks had arrived on the scene, and soon three more joined the dance, which now became more violent, and even temporarily degenerated into a fight as the old cock chased the newcomers from the territory he had established in the clearing. The fight, however, did not last long; the clearing was large and could accommodate several performers without difficulty, for these grouse are gregarious birds that evidently enjoy performing their courtship rituals in groups. And so the dancing, bowing, and posturing, the squeaking, hissing, and resonant rolling calls continued in concert, until the cock who led off the chorus stopped as suddenly as he had begun, shook and folded his feathers, and flew off in search of needed sustenance after his strenuous performance. The others followed suit not long afterward, and soon the clearing once more lay quiet and deserted under the early-morning sun. The next morning at sunrise, and many mornings thereafter, the cocks would dance again, hoping to attract the females with their antics. This they may or may not achieve; although hens usually are drawn to the dance sites and there choose mates, the males sometimes have to search them out.

Observers who have witnessed the courtship ritual of the Black Grouse—the famed "blackcock" of the small-game hunters—agree that this is one of the most impressive and unusual spectacles the spring season in the Eurasian region has to offer. There are those who claim that the performance of the closely related Eurasian Capercaillie—the name comes from the Gaelic word *capullcoille*, whose meaning appears somewhat uncertain—the most famous as well as the largest of this group, surpasses that of the Black Grouse. Others consider the latter a much more accomplished performer;

such preference must remain a matter of individual taste and opinion. There can be no doubt, however, that anyone who has ever seen and heard blackcocks during their courtship display will never forget it. Both the dance itself, with its jumping, pivoting, and prancing steps, and the vocal accompaniment are outstanding performances, and more than compensate for the cock's lack of any highly modified, eye-catching ornamental plumes such as those of the birds described earlier.

The Black Grouse, thus, is a good example of the species whose males use relatively modest physical attributes to their maximum advantage during courtship. Although they all have some distinctive features lacking in the females, these features may amount to no more than different—though not necessarily brilliant—coloring, a larger and heavier body, or both. Other species have erectile tufts, crests or collars, longer or differently shaped tail feathers, or inflatable air sacs in the throat region that make them look bigger and at the same time increase the resonance of their voices. It is the way in which these attributes are utilized to accentuate their acrobatic antics during display, however, as well as the vocalizing that usually accompanies them, that permits the birds to turn their rituals of love into spectacular performances.

Its large size and predominantly blackish coloring lend a special emphasis to the Capercaillie's display. A stocky, heavy bird, the male is about three feet long, or almost twice the length of the brownish, mottled hen. In contrast to Black Grouse, male Capercaillies prefer to display in trees rather than on the ground; their behavior differs in other respects also from that of their close relatives. Whereas blackcocks, for example, are willing to share even their dancing grounds—provided each cock respects the others' individual territories within that area—and seem to enjoy the company of others of their kind throughout the year, the Capercaillie is essentially unsocial, a loner who is jealous of his privacy.

This tendency becomes abnormally strong during courtship, when a cock will furiously attack and drive off any other male that dares to venture within a hundred yards or so of his display area. Capercaillies reportedly suffer from a kind of temporary madness during the breeding season that makes the ordinarily shy and wary birds oblivious to all danger; cocks are known to have attacked even human beings that had wandered into their territories.

For his display, the Capercaillie usually selects a thick horizontal branch not too high up in the tree; dead limbs that offer no obstacles to his movements and permit him to be clearly seen by the hens on the ground are often favored. Strutting back and forth on this perch, wings drooping, tail spread in a fan turkey-fashion, and with his neck arched back so that his beak points skyward, the male gives voice to the distinctive grating, slurring, and hissing sounds that announce his amorous mood to the world. At the height of his performance he seems quite literally deaf to what goes on around him, even to the point of ignoring the sound of gun shots. Hunters invariably sneak up on and bag the bird during its courtship display; at other times of the year it is much too wary and, aided by its normally acute hearing, almost always manages to evade the approach of enemies. Some naturalists believe that the swelling of the neck areas during display may press upon and partially obstruct the bird's ear ducts, so that its sense of hearing is in fact temporarily impaired; combined with the noise it makes, that would account for its deafness. Others, however, are convinced that the Capercaillie's condition is essentially psychological, a result of the trancelike ecstasy that affects this as well as some other male birds during the height of their prenuptial performance.

Be this as it may, the unfortunate result of this striking bird's preoccupation with its courtship ritual has been its virtual extinction over most of its natural range, an expanse that formerly included all the great evergreen forests from Greece and Spain throughout Europe to Scandinavia and Russia. Although able to hold its own easily against the four-footed predators of its range, it could not withstand the depredations by human hunters, who considered the magnificent creature a coveted prize, and mercilessly gunned down the finest, strongest, and healthiest specimens during the height of their sexual excitement when they were easy to approach. Combined with the destruction of much of its natural environment, overhunting resulted in eradicating this large grouse in all but the most inaccessible localities of its range. It is still fairly numerous in parts of Scandinavia and the Siberian taiga; in central and western Europe, however, Capercaillies are found only in private game preserves and some mountain forests, and even there only in small numbers.

The American relatives of the Eurasian grouse all have their own distinctive courtship rites, many of which also are spectacular, even though the largest of the American species does not attain the impressive size and bulk of the Capercaillie. The most outstanding rituals are performed by the Sage Grouse, the Sharp-tailed Grouse, and the Prairie Chicken. All three are today found only in the west-central parts of North America, although the Prairie Chicken formerly occurred in great numbers along the Atlantic seaboard; this eastern variety, known as the Heath Hen, disappeared some 140 years ago as a result of the usual combination of overhunting and environmental encroachment. Like their Eurasian relatives, the American grouse are easy prey for hunters because of the commotion and noise that accompany their courtship rituals, as well as of their affinity for certain display grounds which many of them visit year after year; these habits have resulted in widespread depredations and extinction of local populations.

All three species mentioned earlier are distinguished by externally visible, inflatable air sacs, which serve to increase the reso-

Male Prairie Chicken during the breeding season

nance of the strange booming calls uttered during courtship, and also help to lend the bird a distinctive appearance. The largest of the three is the Sage Grouse, second in size only to the Capercaillie within the entire group; the male reaches a total length of about thirty inches. Ordinarily, this mottled greyish-brown bird, whose plumage seems designed to blend into the background of its native surroundings, does not look particularly impressive. During the

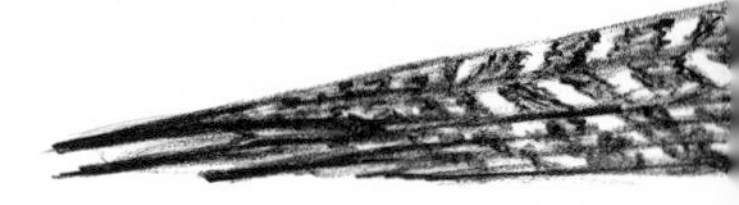

courtship dance, however, it is completely transformed. While the cocks prance, pivot and turn to display themselves from all sides to the hens, they puff out their ornamental ruff of white feathers growing from the loose skin of the neck and breast, so that they seem to be wearing huge white feather boas in which the head all but disappears. In addition, the air sacs are inflated, the tail with its sharply pointed feathers is erected and spread; throughout this display, the peculiar booming sounds accompany the posturing and dancing. Each cock tries to attract and mate with as many hens as will visit his "booming site." In keeping with the habits of most polygamous species, nest-building and rearing of the young are left entirely to the females.

Male of the Sage Grouse

The Prairie Chicken is much smaller and less bulky than the Sage Grouse, attaining only a little over half the latter's size. It also lacks the impressive feather boa, as well as the long pointed tail feathers, having a shorter rounded tail instead. During courtship, however, it makes the best of what it does have; erecting the tail and half-spreading the drooping wings, it inflates the bright-orange air sacs on both sides of the neck until they look like small balloons. At the same time, tufts of erectile blackish feathers growing

just above the bare skin of the air sacs, and normally hanging down over these areas, stand up straight behind the head, thereby forming a backdrop for the also bright-orange "eyebrows," the raised wattles that frame the eyes.

Despite the rather dramatic changes in appearance effected by these features, it is the Prairie Chicken's accompanying dance that provides the real spectacle. Males gather each year at the same booming grounds, where each occupies his own individual terri-

tory. Here at dawn every morning, and then again at dusk, the birds go through their elaborate courtship antics. Head and neck thrust forward, air sacs inflated, tail raised, wings drooping, the cocks utter their peculiar, three-syllable booming calls. At the same time, they stamp their feet, pivot, turn, and run in circles. If a considerable number of males dance together, the sounds made by their stamping may be heard from a distance of up to a hundred feet away; their booming calls, of course, carry much farther.

As many as fifty males may partake in these communal courtship displays, which begin weeks before the females arrive in late March or early April. At that point the crescendo of dancing and booming increases steadily until the birds are ready to mate. Each cock tries to win as many hens as he can entice away from his rivals; the smaller the number of females, the fiercer the competition. Mating, which takes place right there on the booming grounds, ends the brief relationship between the sexes; the females leave immediately to nest and raise their brood.

The Sharp-tailed Grouse, which resembles the Prairie Chicken in both size and general appearance, has a somewhat more northern distribution. The most easily recognizable feature distinguishing it from the other is the central pair of pointed tail feathers extending beyond the rest of the tail which gave the bird its name. The courtship performance is also similar to that of the Prairie Chicken, differing only in details such as the sounds made during display.

All five species of grouse described so far have two characteristics in common: they engage in spectacular displays during courtship, and they are polygamous. In both, they differ considerably from other close relatives, including the Eurasian Hazel Grouse and the North American Spruce Grouse, both of which are monogamous. Although the females of these species incubate the eggs and raise the young alone, the father later joins the family, which stays together as a group until the fall. It is probably not entirely coincidental that the cocks of these monogamous species have fewer and less striking special features distinguishing them from the hens, and that their courtship is much less flamboyant than that of their polygamous relatives.

Among the smaller American grouse species, the one with the most distinctive display is the solitary and monogamous Ruffed Grouse, perhaps the best known of all the American game birds, whose range extends over much of temperate North America. Its name stems from the rufflike patches of feathers on both sides of the male's neck. These patches are spread and a small crest is erected during the courtship performance, whose most outstanding feature is the hollow "drumming" sound made by the cock's wings. This sound has led to much arguing among observers as to how exactly it is produced. Some claimed the grouse beats its wings against the log on which it habitually perches during display; others insisted that the bird beats its wings against the sides of its body. Both assumptions have been proved wrong by slow-motion films, which settled the argument once and for all by showing that the grouse beats nothing but air; it is the rapid motion of the cupped wings that produces the hollow thumping that may end with a sound like a distant roll of thunder.

Vocal performances have an important part in the courtship displays of most grouse; other birds, however, have to rely almost entirely on physical acrobatics in the prenuptial efforts that eventually lead to mating. Renowned for this type of more or less silent display is the Ruff of Eurasia, a bird belonging to the sandpiper group and closely allied to the snipes and woodcocks. The males of this species develop a peculiar collar of long, variegated feathers around the neck during springtime; their coloring is so variable that literally no two birds are alike, and may range all the way from almost black to very light-colored with patterns that include

tints of brown from pale tan to deep cinnamon or maroon. The most unusual characteristic of the Ruff, however, is the behavior that has resulted in its German name, *Kampflaeufer*, which literally means "fighting runner." During spring, males gather in numbers at special display grounds known as *leks*—a name now frequently applied to all avian courtship staging grounds—and there engage in spirited duels. In the course of these fighting dances, the feather collars are raised and spread wide. Although each pair of combatants stab at one another, the duel looks more serious than it is, for the soft bills would do little damage even if they could penetrate the thick protective shield of feathers. Despite the fact that dozens of males may partake in this "social fighting," they always are paired off one against only one other; there never is any ganging up by several on one individual.

During these fighting displays, which take place in almost complete silence, the birds assume the most bizarre poses, quiver all over, and stick their bills into the ground. Although such antics are undoubtedly some form of courtship display, they, like others described earlier, are puzzling in that they often take place in the absence of females. It thus appears that the fighting may serve primarily as a means of establishing a hierarchy among the cocks, rather than as a device to attract the females. This theory is strengthened by the observation that the very sight of another individual with a feather collar, even among Ruffs kept in captivity, excites any full-grown male. Young cocks, on the other hand, who have not yet fully developed collars do not even attempt to enter the main arena.

When the hens, which are known as reeves, arrive at the display site, the males are spurred by their presence to an excess of posturing, quivering, and sticking-bills-in-the-ground. Females choose their mates, and the strongest, most virile cock presumably wins the greatest number of hens, for like most male birds that stage spectacular prenuptial displays accentuating special attributes, the Ruff is polygamous. The fighting arena then becomes the mating site; later on, the females go off on their own to take care of the nesting chores and the rearing of the new generation.

Intriguing as it is, the Ruff's courtship display cannot compare, if only because of the size of the bird involved, with that of the Great Bustard, largest and heaviest of European terrestrial birds. Standing some forty inches tall, this bustard is one of the best examples of how birds with rather ordinary feather colors and patterns can use them to stage an impressive display.

The Great Bustard is a handsome, well-proportioned bird. The sexes are largely similar, but the smaller female lacks the male's chestnut-colored neckband and the whiskerlike tufts of thin grayish feathers that adorn both sides of his head. The mottled tan and brown of the plumage is set off by the white undersides and white patches on the wings. The long, strong, three-toed legs are well adapted for running, which bustards usually prefer to flying. Because of this, the Great Bustard has sometimes been nicknamed "Europe's little ostrich," even though bustards are related to the cranes and plovers rather than to the ratite group that includes the ostrich. Ironically, the Great Bustard's scientific name, *Otis tarda* —from which the term "bustard" was derived—means "slow bird," a name with which anyone watching a bustard running from a pursuer would heartily disagree. Endowed with an acute sense of hearing and excellent eyesight, these tireless runners had little to fear from natural enemies until man decided to decimate their ranks.

Formerly found on plains throughout the entire Eurasian region as well as on the British Isles, the bustard has shared the fate of many large and outstanding birds of those and other parts of the world. Its great shyness and extreme caution long protected it against the more primitive human hunting methods, but modern weapons and devices proved to be more than a match for the bustard. Cultivation of most of the land did the rest, and within one century the Great Bustard became a rare and endangered species, which has been driven from all except a few remote portions of its former range, where it continues to exist even now only because of strict protection.

Owing to the bird's extreme shyness, it has not been easy to discover details about the bustard's habits and behavior in the wild. Although some of its relatives are believed to be polygamous, the Great Bustard as a rule seems not to be; occasionally, however, a strong old cock may win an additional female during battles that sometimes turn into violent and hard-fought duels. Its courtship display, however, is reserved primarily for the female; in her presence, the cock struts and postures, bows, scrapes, and pivots in a

dance that is underlined by the male's truly amazing transformation, during that ritual, from a more or less brownish bird to a predominantly white one.

The bustard begins his astonishing performance by half-spreading his wings, raising his tail, inflating the air pouches in his throat, and erecting his whisker tufts. At the climax of the display, the tail is spread and brought forward until it almost touches the neck. At the same time, the wings are twisted around to expose and erect the white-feathered undersides, "until the bird looks like a huge ball of rumpled up feathers." That is the description given of the bustard's appearance by naturalists F. H. Knowlton and R. Ridgway in their old but very good and accurate book *Birds of the World*. "In this attitude," these observers continue, "it totters and struts about before the female in an exceedingly grotesque manner. It is also very pugnacious at this season, attacking others of its kind and even, it is said, human beings."

The hen alone builds the nest, which is usually located in tall grasses and cleverly hidden from sight. She also attends to the incubating and the early rearing of the chicks. Later on, however, the father takes an active part in leading, guarding, and protecting his offspring until they are old enough to fend for themselves.

Although the bustard is the largest of all the completely terrestrial birds in Europe, it is but a dwarf compared with the African Ostrich. If only because of its tremendous size—a full-grown male may stand eight feet tall and weigh up to 300 pounds—the courtship performance of this bird is uniquely impressive. In addition, the male can—and does—display the pure white, highly modified plumes of wings and tail that contrast strikingly with his otherwise black feather dress; these colors easily distinguish him from the brownish female. It would be logical to assume that the conspicuous plumage of the male is a perfect example of special sexual attributes having evolved in the male as a means of attracting and winning the female in accordance with "natural selection." This assumption would be erroneous, however, for the male's striking plumage has been found to be the neutral, ancestral feather dress of the species, and the undistinguished brown plumage of the female the sexual adaptation. Old, sterile hen ostriches, whose sex hormone production has ceased, frequently begin to develop the showy, black-and-white plumage normally associated with the male. The same change can be achieved by spaying younger females, thereby proving conclusively that the inconspicuous brown feather dress of the hen is in fact the sex-induced change. This, by the way, is true of several other instances of pronounced sexual ornamentation among birds; old hens of the common barnyard variety of chicken sometimes develop the long, distinctive, curved tail feathers of the rooster.

In view of these facts, it may be assumed that the male's behavior, rather than his coloring or plumage, is the most important factor of the courtship performance with which the ostrich woos and wins the female, and his antics during that display seem to bear out this premise. The big, heavy bird puts on an amazing and wonderful show, now strutting, pirouetting, and pivoting, now swaying and spinning at a dizzying pace. At the climax of the performance, the male drops into a crouching position which looks very much as though he is kneeling in supplication before the hen. Underlining this impression are his movements, for he spreads and waves his wings, while his bare neck turns a bright deep pink with excitement. The female looks on apparently unmoved as the male thus seems to grovel before her; since the display eventually leads to mating, however, appearances are evidently misleading.

Because several hens may lay their eggs in a single nest, it is commonly assumed that ostriches are polygamous; this, however, remains to be proved. Be this as it may, the male is a devoted father who not only does most of the incubating but also later on

accepts the larger share of the task of rearing and protecting the young, a trait that is typical of the flightless birds of the ratite group. In species such as the South American rheas, the cock is polygamous but not only takes over all the chores of nest-building, incubating and rearing the young, and is so diligent in pursuit of his duties that he even drives the females away from the nest once he starts incubating.

So far, the species whose acrobatic courtship antics have been described in this chapter are all either large or at least medium-sized birds. This does not mean, however, that small birds cannot stage such shows in order to impress their would-be mates. Many do just that, and none does it better than the manakins, a colorful group of small birds whose males, although lacking special ornamental plumes, are distinguished from the inconspicuously olive-green or -brown, sparrowlike female by their bright and sometimes gaudy colors. In order to properly dazzle the female by both color and movement, manakins engage in acrobatic performances that rival or even surpass those of many larger and more impressively

endowed avian gymnasts. These shows are accompanied by peculiar snapping, rasping, and clacking noises, made by the wings in ways not yet fully understood.

Different species of manakins select different types of stages for their performances. The Yellow-thighed Manakin, a velvety-black bird with golden yellow upper legs and a bright red head—which accounts for its also being called the Redheaded Manakin—

chooses a branch fairly high up in a tree where the foliage will not obstruct its movements, and there performs its bouncing, fluttering dance. Other species, including the Black-and-white Manakin, go to considerably more trouble in order to set a proper stage for their performances. In a manner reminiscent of the Magnificent Bird of Paradise, the male selects a spot where two saplings grow not far from one another, and then removes all leaves and debris in a wide circle around the trees to make sure that nothing is left to impede its freedom of movement. If this in itself is an astonishing achievement for a five-inch bird, the subsequent dance is no less remarkable. Flitting and darting to and fro between, as well as up and down, the two saplings, and producing those peculiar snapping sounds with its wings, the male assumes postures designed to show off its contrasting color patches. Sooner or later, the females are attracted by these antics, and the male mates with as many as he can win, for manakins are highly polygamous.

Dancing on a special stage cleared and prepared just for that purpose is the chosen method of courting for another group of small avian acrobats known as widowbirds, which make their home in the open grasslands and savannas of Africa south of the Sahara. Widowbirds, often also called widow weavers, are a small subfamily of only nine species related to the typical weavers but with very different habits. Males in breeding plumage typically have long black tail feathers, which are believed to have inspired their popular name. Some species display these plumes in hovering, flapping flights during courtship, while others prefer to stage intricate acrobatic performances on the ground as a means of attracting the females.

One of the showiest of the group is the Paradise Widowbird, also known as the Paradise Whydah. In full breeding plumage, the male measures about fifteen inches, but almost two thirds of this length is taken up by the tail, or rather by the greatly lengthened and broadened central pair of tail feathers. The bird's coloring is simple yet striking: a bright red and orange-yellow underside and collar set off the shiny black of the head, back, and long tail. This distinctive appearance adds greatly to the attraction of the male's courtship display.

After selecting a suitable location, the widowbird clears the stage by removing debris and beating down the vegetation within a circle of approximately four feet in diameter, but leaving a single tuft or column of grass standing in the center. Around this centerpiece he then begins to dance, leaping high into the air, prancing and strutting and capering. At the same time, he arches and spreads his long tail feathers in a way the drab, plain-colored females evidently find irresistible. The cock mates with as many hens as he can attract to his dancing ground, for widowbirds are promiscuously polygamous.

The usual ending to this story would be that the females go off on their own after mating to build their nests and raise their families, while the male continues to play around and entice additional hens. In the case of the widowbirds, however, the story has a different twist. Females do not get saddled with the nest-building and incubating chores, for the simple reason that they unload these responsibilities on unwary birds of other species. Like the cuckoo, widowbirds are brood parasites; after depositing her egg in the victim's nest, the female goes her merry way. Unlike the cuckoo and some other brood parasites, however, the young widowbird does not kill any of the legitimate offspring of its foster parents, although the female widowbird may destroy one of the latters' eggs simply to make room for her own. In what is surely one of the most amazing and all but incredible animal true-life stories, the young widowbirds resemble the nestlings of the foster species so closely in every detail of plumage and appearance that even trained ornithologists find it hard to distinguish the two. They also

usually remain with the flocks of the foster species for some time after leaving the nest. How and when the young widowbird finally discovers that it is quite literally a "bird of a different feather" and joins its own kind is one of the countless mysteries of animal behavior that biologists have as yet been unable to solve.

4. *Wooing by Work*

THE RAYS of the sun filtering through the treetops of the New Guinea jungle outlined the hutlike structure on the forest floor with the young sapling rising from its center, and played on the colorful decorations spread out on the carpet of thick green moss that surrounded the hut. A small brownish bird suddenly appeared with an orchid blossom in its beak and added it to one of the neat little piles arranged on the moss. The center pile in front of the hut was the largest; here blue and green iridescent beetle wings glit-terred as the sun touched them. To both sides, heaps of colored pebbles, berries, and fresh flowers made a colorful design on the green moss. The bird circled around to view its masterpiece from a different angle, for this insignificant-looking little fellow was one of New Guinea's famed bowerbirds, the artist-architects of the avian world. As the male, recognizable by his size and plain brown plumage as a Crestless Gardener Bowerbird, picked up several of his ornaments to rearrange the pattern, a female who had been watching from a nearby branch fluttered closer. This immediately spurred the male to a frenzy of activity: grasping one of the blossoms in his beak, he began to disport himself before the female in a manner designed to hold her attention. Half-hopping, half-fluttering around on his moss-carpeted stage, he picked up one colored object after another and displayed them to the female. His earlier attempts had not been successful in attracting a mate; this time, however, the female seemed unable to resist the enticing display. Entering the magic circle of the moss carpet, she joined her suitor in his bower for mating. Shortly thereafter, she flew off to build her nest and raise her brood unaided by the male, who stayed behind to tend the bower he had built and decorated in painstaking work over a period of several weeks.

Although polygamy is suspected in some species of bowerbirds, most apparently mate only once each season. This does not deter the males from tending the bower as assiduously after it has served its initial purpose as before. Wilted flowers are replaced with fresh ones; damage to the structure is repaired; the ornaments are kept in order. To date, the best available explanation of these activities is simply that the males enjoy them.

Of all the birds whose courtship includes "construction work," that of the bowerbirds is not only the most intricate but also the most fascinating, because it does not involve nest-building, and also because of the esthetic, artistic factors involved. The first naturalists to discover the incredible feats of these birds, some of which make their own paint and use tools to apply it to their bowers, waxed enthusiastic about the feathered artists, speculating that in these birds man had found proof of the basic natural wellspring from which human artistic talent also derives. This concept was soon replaced by the mechanistic view of the functionalist biologists, who denied any and all conscious esthetic or artistic elements in the activities of these birds; the bower-building was considered as nothing more than an instinct-activated means to supplement the lack of outstanding plumage in the male and help attract the female.

In the meantime, the pendulum of biological opinion has swung to somewhere in the center between the two views. On one hand, it has been proven that the bowerbirds' artistry, and even the selection of the ornaments with which they decorate their bowers, is indeed bound by instinct which impels the males of one species to always select the same colors, always use the same tools for painting the walls, showing that the little artists have very little freedom to experiment in their "artistic" endeavors. On the other hand, it also has been proved that not all individuals of the "painters" species do in fact paint the walls of their bowers, nor does such a lack of extra adornment seem to have any adverse effect on their ability to attract a mate. Beyond that, careful observation has shown without a doubt that these birds' activities far exceed the narrow confines of a sex-oriented display serving reproduction only. The way in which the males tend their bowers long after mating—replacing wilted flowers with fresh ones, rearranging the colored objects so they will catch as much light as possible, while removing all debris—indicates that they get a heightened sense of well-being from these activities. We also have to admit that the color patterns and arrangements of the ornaments in many cases are evidence of a sensitive awareness of visual attraction that coincides remarkably with our own. Despite all protestations about blind instinct, the bird that woos its mate with a freshly picked orchid in its beak is so uniquely reminiscent of human courtship practices that no one can fail to be impressed.

Bowerbirds are a small family of only nineteen species restricted to New Guinea and northern Australia. They have been found to be closely related to the birds of paradise, despite their lack of ornamental plumes. The two most primitive members of this group are known as catbirds because of their peculiar mewing calls. The sexes are similar in both species, the males build no bower, and practically nothing is known about their courtship and breeding habits.

Next in line are the so-called stage builders, of which there are several species. The courtship practices of these birds remind the observer somewhat of such members of the Paradisaeidae as the Six-wired Bird of Paradise, which also clears a circular "stage" on the forest floor for its courtship performance. The stage-building bowerbirds cover the floor with layers of leaves, sometimes selecting only those of a single type of plant. Several species leave a single sapling standing in the center of the stage, and pile twigs and debris around the stem of this small tree in a column that may attain a height of three feet. They then display and sing while hopping from the top of the column to the stage floor and back. Wilting leaves are usually replaced with soft fresh ones.

The remaining species of bowerbirds are grouped into two categories by naturalists, the first of which comprises the so-called maypole builders, whose bowers are large and bulky. So large, in fact, are the hutlike arbors built by one species that the first naturalists to see them believed them to be playhouses erected by the aborigines for their children. It came as quite a shock to find out that structures towering up to nine feet high were the work of a nine-inch Golden Bowerbird.

The handsome male of this species first clears the ground around a young tree that has a "twin"—another similar or slightly smaller sapling growing not more than a few feet away. These small trees will become the supporting centers of the bird's structure. Now the male piles up sticks, twigs, and other debris around the stem of the larger sapling until this column reaches a height of several feet. The little architect then assembles a similar but smaller pile around the base of the other sapling. The single most astonishing feat is the construction of the display perch; this rather thick horizontal branch bridges the gap between the two columns, and is then decorated with mosses, lichens, and other plants. From this lofty stage the male woos the female with both his display antics and his varied song repertoire, for the many-talented bowerbirds are skilled and versatile mimics.

The typical maypole builders forego the elevated display perch and devote more attention to the ground around the bower, which usually has a hut- or tepee-like appearance because of the thatched roof provided by the feathered architects. The supporting center of these arbors usually is a single column of twigs and debris piled around the base of a small tree; the roof, which is so well-constructed that it can withstand heavy rains, reaches to the ground in the back and on the sides, but leaves open the front. The "front yard" is then cleared of all debris and decorated. Especially the species known as "gardeners"—such as the Crestless Gardener described at the beginning of this chapter—devote a great amount of work to beautifying their bower entrance. Methodically laying down a semicircle of thick moss, the birds arrange the colored ornaments in definite patterns on this carpet, and frequently even decorate the central column. The German naturalist Heinz Sielmann, who made several filming expeditions to Australia, observed during one of his long waits in specially constructed blinds that an Orange-crested Gardener arranged his ornaments according to color, separating a collection of iridescent blue beetle wings from a group of snail shell fragments by placing a line of yellow flowers in the center. Sielmann was amazed at how the bird, like any human artist, time and again retreated to view the overall effect from a distance, and then proceeded to rearrange several of the ornaments

for a color pattern that was evidently more satisfactory.

Considering the magnitude of both the construction and decoration tasks, it is clear that each of the more elaborate arbors is the product of weeks of work by creatures that have only their beaks and feet to work with. A single bower has been found to contain some 3,000 individual sticks and twigs as well as several hundred ornaments. Some bowerbirds return each year to the same bower, repairing and adding to the structure; others evidently build a new one for each mating season. Observations seem to indicate that the feathered architects learn from experience, for younger bowerbirds construct much less intricate and well-adorned bowers than older males.

Although the arbors of the final group of bowerbirds, the so-called avenue builders lack the thatched roof of their maypole-building relatives, biologists consider the half-dozen or so species of the former to be the most advanced of the entire group. The avenue builders use no saplings as supports for their structures; instead, as a foundation for their corridorlike bowers, they painstakingly lay down a mat, sometimes several inches thick and made of sticks and twigs. This mat is a perfect support for the bower walls, which consist of long sticks pushed more or less upright and close together into the mat in parallel rows. The resulting "avenues," which are wide enough to permit the bird to pass through without brushing its wings against the walls, always run in a north-south direction. This fact puzzled naturalists for some time, until they realized that the main display area, which is located just outside the northern entrance, gets the sun all day long during that particular time of the year in the southern hemisphere, which in turn means that the ornaments are illuminated from sunrise to sundown.

The bowers of the avenue builders vary considerably with the species, and may range from the relatively simple structure of the striking black-and-yellow Regent Bowerbird to the elaborate arbor of the drab-looking grayish Lauterbach's Bowerbird. The latter adds a third and fourth wall at opposite ends of the avenue to create a kind of open-cornered enclosure. There seems to be a definite correlation between the birds' own coloring and the ornateness of their bowers, almost as though the more drab-looking species feel compelled to compensate for their lack of colorful plumage by building more elaborate structures.

Each species of avenue builder has its own favorite color combinations. Lauterbach's likes blue and red; the handsome blue-black Satin Bowerbird shuns red hues but prizes blue and yellowish tints. Both the Great Gray Bowerbird and the closely related Spotted Bowerbird, on the other hand, also like red but concentrate mostly on white, pale yellow, and silver objects; these birds are especially keen on anything that glitters and will pick up such items as broken glass or small shiny metal utensils. The urge to collect drives these birds to steal any small glittering or sparkling object they can carry, and they regularly raid any human habitations located in the vicinity of their bowers. One Australian who had left his car keys in the ignition found them missing when he returned. Knowing that the arbor of a Spotted Bowerbird was located not far away, he knew exactly where to look for his property; he found his car keys hanging at the bower entrance.

Modern civilization has provided these bowerbirds with the opportunity to add such other unusual items as bottle tops, hair curlers, electrical cords, and metal buttons to the snail shells, pebble, and bleached animal bones with which they normally decorate their arbors. One Great Gray Bowerbird was observed displaying proudly with a shiny tin toy colander in its beak; a tiny plastic red Santa was another's prized possession. Because of this overwhelming urge to collect, Australians sometimes call a "bowerbird" anyone who accumulates odds and ends.

The discovery, about fifty years ago, of the fact that a few species of bowerbirds paint the walls of their arbors caused great excitement. To find a bird employing a tool in order to engage in a sophisticated and nonvital activity—it has been proved that the painting of the bower walls is not essential for attracting the female—was an epoch-making discovery. Although the urge to paint is an inherited trait limited to the use of certain colors and "brush" materials—the handsome Regent Bowerbird, for example, applies only greenish-yellow tints with the help of a wad of leaves —there is no denying that the activity is unique precisely because it

is so unnecessary. The only other species of bird known to employ a tool in the conventional sense is one of Darwin's famous Galapagos finches, which uses a long thorn to pry insects from under the bark or out of crevices. Although this is an astonishing feat, it is more readily understandable because animals in general display their maximum capacity for ingenuity in the pursuit of food.

The "paints" used by the bowerbirds are usually either the juice of certain berries or other plant parts, which the birds crush, or charcoal mashed and mixed with saliva to produce a thick paste. The "brushes" may consist of a wad of bark fiber or of crushed leaves, which the birds use to apply bands of dark gray, blue, or greenish-yellow color to the insides of the bower walls. As mentioned earlier, some males of the three "painter" species evidently feel no urge to decorate their bowers in this manner; it therefore appears that the birds have some freedom of choice in that particular area of their artistic endeavors.

The earlier-mentioned Satin Bowerbird is perhaps one of the best-known "painter" species. In the past, the bird decorated its bower walls with bands of both charcoal black and dark berry blue. Human civilization, however, supplied the bird with a new, brighter, and easier source of color. Since blue is the Satin Bowerbird's favorite hue, males will pick up any blue object they can carry, and those living near human habitations found the small bags containing laundry blue a ready-made prize. One exasperated Australian housewife lost so many that she blamed and punished her small son for taking them until she found out a bowerbird had been the culprit.

It did not take long for the birds to discover that the contents, mixed with saliva, made a bright blue paint far superior to the charcoal and berry juice they had been using, and soon the bowers of birds living within flying distance of human homes were distinguished by bands of laundry blue.

The Satin Bowerbird, which is found in eastern Australia, is among the largest of the nineteen species of this group. Its bower, however, is neither as large nor as richly decorated as that of the much more drab-looking Lauterbach's Bowerbird, whose arbors have in some instances yielded more than a thousand colored pebbles and berries. The Satin Bowerbird never goes to quite those lengths in decorating its bower, although individuals, especially older males, may collect a considerable number of blue and yellow items such as feathers, flowers, and snail shells—up to one hundred blue feathers and flowers have been counted in one of these birds' bowers. It is common knowledge in eastern Australia that gardening enthusiasts who raise blue flowers in the vicinity of a Satin Bowerbird's territory have to expect ruthless raids; as soon as a blossom opens, the male is sure to be along, snip it off, and carry it back to its showplace.

From observations of the Satin Bowerbird after completion of its arbor, the entire sequence of courtship display and mating activities has become known. With all the ornaments arranged to his satisfaction, the male begins to call in a peculiar manner, and shows off his shiny blue-back plumage while fluttering and hopping around the bower and at the same time picking up some of the colored objects here and there. These antics may attract a female within a very short time; otherwise he keeps on displaying until a prospective bride appears and shows her interest by playing about the bower entrance for a little while. Finally, she enters the bower, thereby encouraging the male to perform even more energetically while chattering and churring continuously. His dance at the climax includes various strange postures where he stands with lifted tail, lowered head, and bill pointing to the ground. As he gets more excited, he spreads his wings so the female can see the glossy deep blue of his feathers, and repeatedly picks up and tosses about blue and yellow objects.

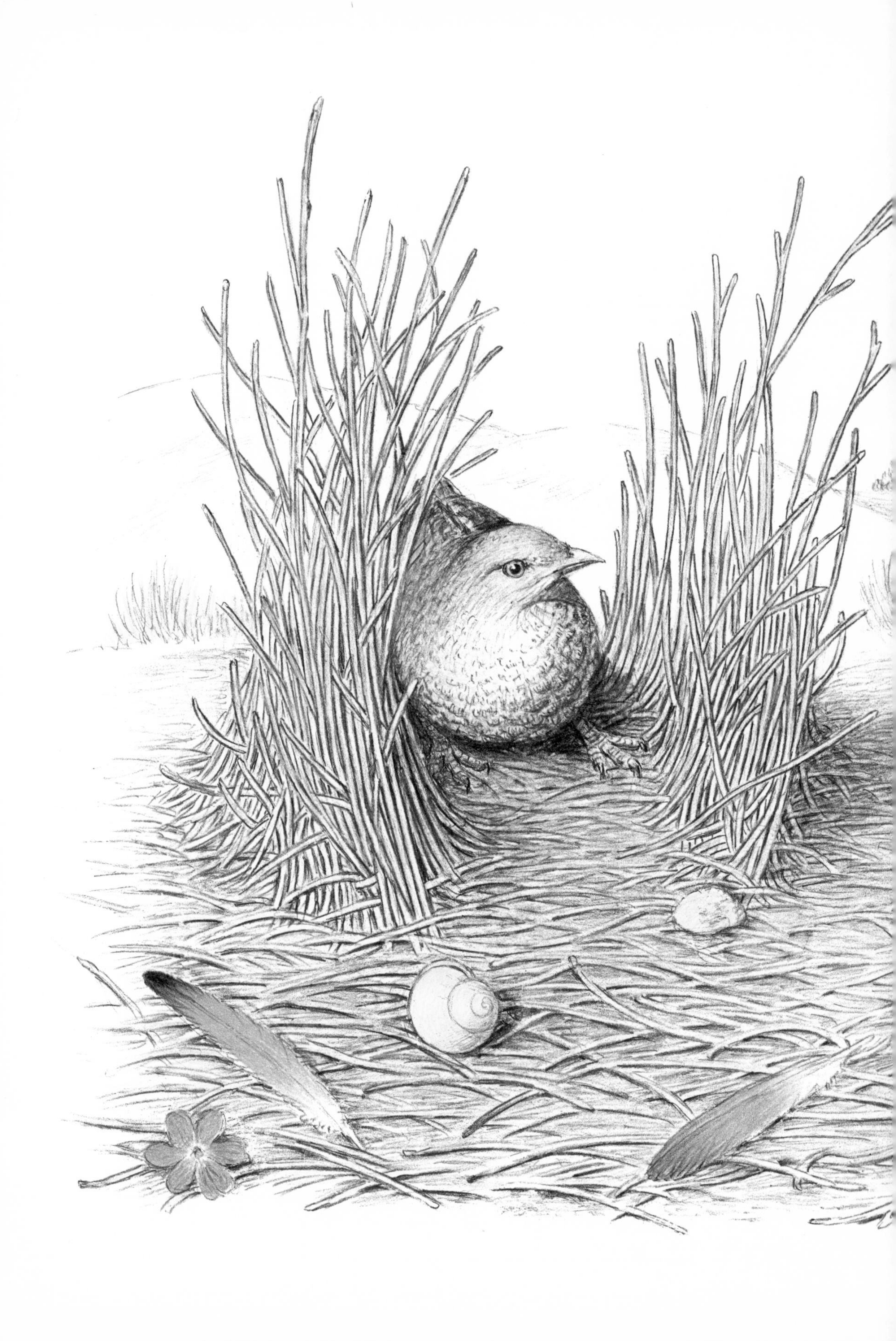

The display may continue for half an hour or more before the male also enters the bower, where mating takes place. In a complete turnabout that has long puzzled naturalists and is still not fully understood, the male then drives his newly won mate from the bower in what looks like a burst of rage. The best explanation to date of this behavior is that the bower, having served its purpose of attracting and winning the female, once again becomes the male's exclusive, prized personal possession, in which no one else is

tolerated. This theory is strengthened by the fact that the Satin Bowerbird, like others of its kind, does not lose interest in its "handiwork" once mating is over. In the weeks that follow, during which the female builds her small nest and raises the young by herself at another location, her mate continues to repair, tend, and add to his bower as devotedly as before.

Bowerbirds thus furnish perhaps the most striking proof of how the traits and talents of a species, although utilized through the influence of sex hormones for the important task of winning a female and thereby insuring the propagation of the species, may result in complex activities far exceeding the needs of mere sexual attraction. Observations bear out the fact that the male's sophisticated work is largely unnecessary for securing a mate. One naturalist secreted in a blind watched, fascinated, as a female Satin Bowerbird slipped into the crude, poorly adorned bower of a young and still greenish-colored male, while ignoring the impressive, richly decorated arbor of a resplendent blue-black older male who had also been wooing her. Retribution for this affront was promptly exacted by the spurned suitor, who destroyed his younger rival's bower in a burst of jealous rage. The young male meekly accepted this punishment, respecting the older bird's seniority, and the female also apparently saw the errors of her ways, for she eventually entered the dominant bird's bower for mating.

The fact that their prenuptial "construction work" is not even remotely connected with nest-building makes bowerbirds unique. The males of all other species that engage in such labor as an essential part of their courtship devote it to preparing a nest for the future offspring, and then offering this nest to the female as an enticement to mate. Although the great majority of birds' nests are built after mating, either by the female alone or by both sexes in a joint effort, there are some species of widely varying size, appearance, and living habits whose males include the preparation of a nesting site as part of their courtship activities. These male-built nests may range from rather crude piles of sticks or twigs to some of the most finely built and clever structures fashioned by birds.

Among the more untidy of such nests are those of the frigate, or man-o'-war birds. These feathered pirates of the seas are the most graceful and accomplished aerial acrobats, and their wingspread of six feet or more exceeds that of any other bird of comparable size and weight. Watching frigates in the air is an esthetic delight as they soar, sail, and maneuver with effortless grace. On the ground, however, they are the exact opposite; even more awkward than their relatives the boobies (popularly known as "goony birds"), they have great difficulty moving about on level ground, and avoid it whenever possible. Also in contrast to all their relatives, and despite their fully webbed feet, they do not like to swim and quickly become waterlogged if forced to do so. When they are not in the air, they prefer to roost in trees and bushes from which they can take to their wings most easily.

Frigates have a predominantly black plumage; the male of the Magnificent Frigate Bird is distinguished by a normally orange throat pouch, the female by a white breast. Beautiful as they are in flight, they do not have a personality that could be described as attractive. Pilfering from other birds is a way of life for these feathered buccaneers. They harass boobies, gannets, and gulls into dropping their catch, and swoop down to snatch any unprotected eggs or chicks, including those of their own kind. Food is not the only thing they steal; during the breeding period, nesting material is ruthlessly pilfered from neighboring nests if the legitimate owners are not watchful.

When the male gets ready to court, he undergoes a color change that announces his mood, for the normally orange hue of his throat pouch turns a fiery shade of red. He now begins to collect sticks

and twigs, which, in a time- and labor-saving effort, he attempts to steal from other nests if at all possible. Staking out his territory with his roughly assembled platform, he squats on the future nest and expands his pouch until it resembles a red balloon. Displaying this courtship signal to an interested female calls for various weird poses, including one in which he throws his head back and points his bill skyward, permitting the pouch to balloon out to its fullest capacity. As soon as a female decides to accept this fiery proposal together with the nesting foundation the male has prepared, she gathers additional sticks and twigs and brings them to the male, who adds them to the nest until the bulky, untidy structure has reached the point where it satisfies the demands of both parents. As soon as the single egg is laid, the male's pouch once more fades to its normal orange color, and he stops displaying. His work, however, is not over, for from that time on both parents have to continuously guard, first the egg and then the chick, against attacks by their cannibalistic neighbors, who—like they themselves—think nothing of making a quick snack of any unattended egg or nestling.

A much more gentle and pleasant personality is displayed by the Adélie Penguin of Antarctica, whose courtship activities also include the male's responsibility for providing nesting material. Together with the Emperor Penguin, the largest of the entire group, the Adélie is restricted to the frozen wastelands of the Antarctic continent. Whereas the Emperor, however, builds no nest at all but keeps its single egg—and later the chick—safe and warm in a special feather pouch between the parent's legs, the Adélie Penguin does build a nest of sorts, even though the only movable material in its bleak homeland consists of pebbles.

The courtship of all penguins, including the Adélie, features much mutual bowing and posturing, but as in many other monogamous species, especially those that nest in colonies, these rituals are widely used social greeting ceremonies not confined to the

breeding season. Research into the Adélie's social and family life has yielded a number of astonishing facts, including the apparent ability of these penguins to recognize their young by their facial characteristics, an all but incredible feat considering the hundreds of nestlings milling about in the breeding colony.

Males start collecting stones at the nesting site even before the females arrive. Offering an especially enticing-looking pebble to the female as a sort of gift is a common practice; among birds that breed for the first time, acceptance of this gift by the female apparently signals her willingness to accept the suitor also. She then takes possession of the nest and rearranges the stones which her mate continues to furnish. Male penguins habitually steal pebbles from nearby nests when the owners are not looking; although this may be somewhat less than neighborly behavior, such convenient sources of nesting material prove irresistible. Completing the nest is quite a task, not only because stones of the right size are not always easy to come by, but also because female penguins may frequently prove capricious, throwing out some stones for no discernible reason. This means that the male has to make additional "collecting trips" until the nest finally satisfies his mate. Even so, addition of pebbles to the nest usually continues long after the female has started incubating.

Although the courting males of species such as the frigate and the penguin may invest considerable time and labor in providing nesting materials, the resulting structures are hardly masterpieces. The exact opposite is true of nests built as courtship offerings by the males of certain small songbirds; these species must be rated among the most expert weavers in the entire animal kingdom, and their creations as outstanding examples of avian artistry.

Prominent among these tiny "working suitors" are the few species known as penduline titmice, so called because of their bag-like hanging nests. These Old World residents are related to the well-known and more common titmice of backyards and gardens, such as the Great Tit of Europe and the Black-capped Chickadee of North America.

Even though both these groups are placed in the same family, and share such characteristics as small size and lively behavior, their nests are completely different. The common titmice are cavity nesters, and either utilize ready-made locations such as abandoned woodpeckers holes, or else fashion their own in soft, decayed wood. The penduline titmice, on the other hand, choose to suspend their nests from the thin outer branches of trees or tall bushes.

A typical member of this group is the Eurasian Penduline Titmouse, a cheerful little mite of only a little over four inches long, with a chestnut back, pale, rose-flushed underparts, and a conspicuous black mask. Both sexes have the same coloring.

The male starts to construct his nest in the spring, anchoring it to twigs of such trees as willows or poplars, usually near water in a boggy or marshy environment. Into a framework of strong plant fibers, the little artisan knots and weaves a variety of finer plant material to fashion a roughly slipper-shaped bag. These nests are of such superior construction that children in Austria are reported to have occasionally used them as footwear—an unusual practice made possible only by the great elasticity of the small, tube-shaped entrance hole, which can be stretched very much like the neck of a turtleneck sweater.

As his task nears completion, the male looks around for a female who likes his masterpiece well enough to accept it as her nursery and him as her mate. She then finishes the interior by lining it with fine, soft plant down, and starts her brood. This leaves him free to build another nest and search for another mate. Since each structure takes from three to four weeks to complete, the male is kept busy for a good part of the breeding season. His brand of polygamy thus is not the carefree kind of many other birds.

Certain other small songbirds have a somewhat different approach to the problem of courtship-by-nestbuilding. Species such as the Blackcap, an Old World warbler, and the familiar wrens construct several nests in succession, and only then take a female on a guided tour to let her pick out the one she likes. The common Winter Wren, which occurs widely all over the Northern Hemisphere, has been closely observed during these activities. Of the several nests he prepares each spring, one is invariably much more

carefully constructed than the others, and it seems that the female just as invariably chooses that one. The male may seek out another female and show her the rest of his nests; if one of them is reasonably well built, he may succeed in winning another mate. Female wrens are discriminating, however, and a male who is careless and builds poorly constructed nests may be left without a mate.

As soon as the female takes possession of what is to be the future nursery, she, like the female titmouse, adds some soft plant fibers as a lining of the interior, lays her eggs, and starts incubating. This she does by herself, although the male usually helps to feed the young.

Despite extensive studies of more than two dozen species of wrens, with interest focused mainly on their courtship, social, and family life, much still remains to be learned about these unusual birds. What has emerged from such studies, however, is clear evidence of some traits and characteristics that set this group apart from its closest relatives. Some species of wrens, many of which are wonderful vocalists, have a composite song performed by both male and female especially during courtship, with the male singing the first part, and the female immediately joining in with the second. Also remarkable is the strong family bond existing between the various members and displayed in the tendency to share in the tasks of housekeeping. Thus the young of the first brood often help their parents to feed the young of the second generation, resulting in the unique spectacle of older birds feeding their younger brothers and sisters.

Sleeping habits also vary among the species; wrens may sleep singly, in pairs, or in family groups. Many facets of the wrens' courtship and social behavior remain puzzling to scientists, and especially the reasons for the excess number of "cock's nests" are still incompletely understood. Evidently an integral and important part of courtship display among wrens, the unused structures seem to represent the male's overwhelming urge to build. Sometimes one of them serves a secondary purpose when the male uses it as a bedroom, for wrens prefer to sleep tucked away in some nook or hole where they feel protected during the night.

The most sophisticated nests built by males as part of their courtship activities are without a doubt those produced by certain species of weaver birds. The male's courtship consists of very little more than offering a finely and newly woven nest to the female. Even the courtship display of those males with brightly colored plumage seems designed mainly to focus attention on the nest.

Weavers, which are grouped in several families and are found in both Africa and Asia, have long been famous for the artistry with which many of them fashion their nests, as well as for their social behavior. Most of these birds are highly gregarious, and some build huge, communal "apartment-house" nests with many occupants under a single roof. Even those that prefer to construct individual dwellings like to aggregate in colonies that may number several dozen couples.

The breeding season of such species as Cassin's Weaver or the African Village Weaver begins typically with a group of males selecting a suitable tree, and each then choosing a site in that tree for its nest. These choices are made with considerable care. The trees selected are usually thorny plants armed with sharp, hard spikes that discourage predators from climbing around in them, and the nesting branches are too weak to support the weight of any animal likely to be a threat. Because of these precautions, weavers make no attempt to hide their nests.

The intricate weaving task required for each structure is a labor of many hours in which the male uses strips of palm fronds and grasses to fashion the flask- or retort-shaped nests, which have an entrance tube and a brood chamber. The bird loops and knots and tucks loose ends into the fabric much like a human craftsman engaged in weaving; it holds the strands in its feet while pushing and

pulling with its beak, until the nest is completed except for the "interior decoration," which the male weaver leaves to his mate.

The females arrive about the time the first set of nests are finished. Each male now tries to attract a bride by displaying at the nest entrance, fluttering around and chattering excitedly. The Village Weaver, whose breeding colors are bright golden yellow and orange set off by a black mask and blackish wings, shows off these hues while clinging to the entrance of his freshly built and still-green nest. Should a male fail to find a mate before the plant fibers worked into his nest begin to dry out and turn brown, he destroys the structure and builds a new one. Should a female choose to accept him and his nest, however, he immediately starts to build another in order to woo a second mate, for these weavers are polygamous. If females are plentiful, the males may go on to construct a third and fourth nest. The quality of the structures may vary considerably, because older males are generally much more skillful and accomplished weavers than the younger birds. The fact that older birds are usually better "craftsmen" has been noticed in many other instances; although partly due to a slow maturing of the weaving ability, it also reflects the learning experience gathered by the older individuals. In this respect, birds differ sharply from other expert weavers of the animal kingdom; the weaving talents of spiders, for example, are rigidly controlled by inherited instincts, so that a spider's first web is as good as its last, regardles of how many it contructs in between.

In all the preceding instances of avian "working suitors," with the sole exception of the unique bowerbirds, the males have to prove their ability to provide not only a breeding territory but also a suitable home—or at least the materials for it—as part of their courtship and as a prerequisite for winning a mate. "No home, no wife" seems to be the rule for all of these species. To the female is reserved the privilege of selecting or rejecting the nest offered to her and, by implication, the male who offers it. The female's choice thus centers on the male's performance as a provider rather than on physical attributes such as size and appearance, which in turn means that he is actively involved in the all-important preparations for raising a new generation. Most of these males, despite the polygamy some of them practice, also help to feed the brood later on. Such mutual participation points the way to those species in which the pair bond is strong and often lasts for life, and who not only partake equally in courtship rituals, but also share the duties of caring for and raising the young, thereby coming as close to achieving a true family life as is possible for birds.

5. *Mutual Courtship*

THE TWO STATELY birds with their distinctive powder-puff headdresses of thin, stiff feathers moved in a synchronized rhythm of postures. They bowed and leaped, spread their wings, and stretched their long necks to point their bills skyward in a strange dance ritual whose every deliberate movement clearly had a very special meaning for both partners. Both Crowned Cranes stood about three feet tall, wore the identical crests of yellowish, black-tipped feathers, and had the same overall coloring. Their white cheeks were offset strikingly by the sooty black crown above and the red wattles of the neck below. The predominantly bluish gray of the plumage, combined with the brown of the tail and the pale yellow of the primary feathers, blended into a pleasant design of subtle hues.

After the birds stopped their dancing, they remained standing quietly facing one another. The slightly larger individual now made the first move by reaching over to nibble its partner's bill. The other responded in kind; standing there face to face, their heads touching, the two cranes presented the classic picture of a couple that belongs together and knows it. An observer could have deduced only from one of the birds' somewhat submissive posture that it was a female, for cranes are typical examples of the numerous bird species that have no outward distinguishing sexual characteristics. This similarity in the appearance of the sexes most often heralds a strong pair bond and full partnership between male and female.

The large, stately cranes are a small but ancient group whose members have been admired—and at times even revered—in every part of the world where they occur. Unfortunately, this did not help them much in the past; although they are now rigorously protected almost everywhere they are found, hunters ruthlessly slaughtered them in earlier years. In addition, some species need such large territories in order to flourish that the steady expansion of the human population has perhaps contributed more to their decline than the overhunting of bygone years. In a few instances, such as that of the magnificent North American Whooping Crane, stringent protection and untiring efforts by conservationists have apparently been successful in reversing the alarming trend that brought the total Whooping Crane population to an all-time low of only twenty-three birds in 1941.

One of the by-products of the extensive observations that were part of these conservation efforts was a detailed knowledge of the whooper's habits. This record of how the big birds live, feed, court, and breed provides fascinating glimpses of the family life of a shy and ordinarily inaccessible species.

It had long been known that all cranes engage in spectacular dances. Although these dances are an integral part of the birds' courtship activities, they also take place at other times of the year and among mixed groups, evidently as a form of social entertainment or simply as an expression of their exuberance and *joie de vivre*. During the breeding season, these dances become prenuptial ceremonies between male and female preparatory to the serious task of raising a new generation.

Filmed sequences of the courtship ritual of the Whooping Crane, taken from carefully hidden blinds, have captured the various poses, postures, and movements that distinguish this dance. The male begins by approaching the female and solemnly bowing to her. If she responds, the couple then perform a series of synchronized leaps, steps, and stances that turn the dance into an awkward yet strangely graceful ballet, which typically takes place in the shallow water near the edge of the marsh in which the whoopers make their home. At the climax of the performance, the male in his exuberance may leap clear over the female.

The dances of other cranes are very similar. Although the sequences and individual postures may vary with the species, they are all alike in that the sexes are equal partners in the dance and that the ceremony is distinguished by an exuberance that seems strange for such stately and normally rather slow-moving birds.

Like others of their kind, Whooping Cranes have a very strong pair bond. Strictly monogamous, the mated couple stay together and faithfully share all responsibilities for raising and protecting the young, which remain with their parents until their dark juvenile plumage begins to change into the adult's white feather dress. Shy, wary birds, these cranes are individualists who like their privacy; a single family needs a territory of approximately four hundred acres, and may be driven off by nothing more than the realization that the territory no longer offers them freedom from intrusion. This reticence is not timidity, however; Whooping Cranes have a fierce, untamed nature, and the male can be a pugnacious and fearless fighter. The usually slightly smaller female is much more gentle.

Establishing suitable refuges for another large, beautiful, and exclusively North American bird has been much easier and more

rapidly successful because this species, the renowned Trumpeter Swan, has proved willing to change its migration pattern and adapt itself to available refuges. The Trumpeter, larger and more striking than the Whistling Swan, the only other native North American species of swan, was fairly common in Colonial times, but was hunted almost to extinction in the nineteenth century. Only a few dozen pairs of the big white birds survived in the United States at the turn of this century, all of them in the protected region of the Yellowstone Park. In 1935, a count turned up a total of only thirty-five Trumpeters, but since then the number has been on the increase thanks to determined conservation efforts plus the swan's ability to change its habits. Today there are some one thousand Trumpeters, most of them in the refuge of Red Rock Lake in Montana, where the National Wildlife Service continues to watch over them and provide food when necessary. These sustained efforts have not only preserved the bird for future generations, but also furnished proof that such determined action can be successful in saving endangered species.

All swans are monogamous and are believed to mate for life. The habits of the Mute Swan, the most common Eurasian species, have been thoroughly studied, because this bird has been more or less domesticated in Europe for centuries. In the past, swans were used for food, and roast swan was considered a delicacy. Today, however, the beautiful white birds are kept for esthetic rather than economic reasons, and grace many lakes and large ponds in public and private parks and estates, where they often build their bulky nests and raise their young.

Male swans are faithful mates and fathers. The Mute Swan especially is renowned as a fearless defender of his mate and family. Incensed over an intruder venturing too close to his nest, a cob can be a formidable adversary, capable of inflicting serious injuries even to large animals, including humans; one blow by a swan's powerful wing can easily break a man's arm. Smaller predators risk their lives by trying to rob a nest. Even a large dog would be ill-advised to challenge a brooding swan.

Although the extremely shy Trumpeter Swan is difficult to approach and observe, a few naturalists have been able, by exercising great care and hiding in well-camouflaged blinds, to record their courtship habits and family life.

By the time they are ready to breed, Trumpeters have already been paired off for some time, for like the closely related geese, swans become "engaged" before they are sexually mature. As the breeding season approaches, paired couples of trumpeter swans have been observed going through a graceful ceremony that takes place entirely on the water. Here, then, is the original *Swan Lake* ballet as the two large, pure white birds, their necks curving gracefully and their wings half-spread, face one another in their prenuptial performance, during which they may rise out of the water and almost literally walk on it, and then again swim about circling while still facing each other, heads bent and close together.

Almost everything that applies to the swans' family life is true also of wild geese. The handsome, appealing Canada Goose, probably the best known of North American species, is as fine an example of a faithful mate and devoted parent as one can find among birds. Geese also tend to pair off before they are sexually mature and capable of mating; the partner selected at that time is chosen for life. The pair bond among Canada Geese is so strong that they do indeed stay together "until death does them part"; when that happens, the survivor often shows definite signs of grieving over his or her lost mate, and may delay choosing another one. In at least one recorded case, it was reliably reported that a Canada gander who had lost his mate remained near the place where she had died, refused food, and soon afterward was also found dead; an examination disclosed no discernible physical cause of death. It

would have been said of humans that the survivor died of a broken heart; to say something of that kind about animals would be indulging in anthropomorphism (the ascription of human characteristics to animals) which to biologists is a major sin indeed. Yet we do not—and may not ever—know exactly what goes on in the mind of an animal. There is considerable evidence that the lack of the will to go on living following the loss of a close companion may afflict certain animals even as it does humans.

The courtship of the Canada Goose, as well as that of other species in its group, involves ceremonial performances in which both partners stand side by side, utter their honking calls, and curve their necks so as to brush against one another. As mentioned earlier, the couples already are paired off before they are sexually mature; the courtship ritual serves only to seal the lifetime bond between the partners. Unfaithfulness or "divorce" is apparently unknown among Canada Geese.

Although swans, geese, and ducks are all grouped into a single family, there exists such great diversity in size, appearance, and habits among the approximately 150 species that they have been separated into two subfamilies, one of which includes the swans, geese, and certain gooselike ducks, while the other embraces the more typical ducks. Next to their generally much smaller size and shorter legs and necks, the most immediately apparent characteristic distinguishing ducks from geese is the sex-related difference in color patterns. Drakes are almost always more brightly colored than the females; in some species, such as the beautiful North American Wood and the Asian Mandarin Ducks, the difference is striking and includes some specialized feather formation as well. One of the most familiar wild ducks is the male Mallard with his bright green iridescent head, white neck band, and cinnamon breast; in comparison, the brownish female looks drab indeed. Although the males of many species wear their distinctive plumage throughout the year, for others it is seasonal. Once breeding time is over, the brilliant colors of species such as the wood duck are replaced by much duller hues.

Like other waterfowl, ducks are monogamous, but both their courtship and breeding habits differ considerably from those of their larger relatives. During courtship, there is a good deal of rivalry and competition among males; the female finally chooses the one she prefers. The males of most species go off on their own as soon as the female starts incubating, thus leaving to her the chores of raising and protecting the young. Only after the offspring are half-grown does the father rejoin his family, which now flocks together with other such family groups.

The courtship of the small, chunky Ruddy Duck of North America has been observed and recorded in detail. The spring molt replaces the gray of the male's winter plumage with rusty red, offset by his black cap and white cheeks, but most of all by his bright blue bill. Courtship takes place entirely on the water, and begins when a male decides to pursue a female. She dashes away across the surface with a sharp cry, which usually is the signal for other males to join the chase. She now begins to dive and dash alternately, always followed or surrounded by two or more males. Finally, she makes a show of threatening one of them. He apparently interprets this as a sign of her preference, and begins to swim and circle about in front of her with sharply cocked tail. If she thereupon accepts him, they both swim away together and are no longer pursued by the other males. Ruddy Ducks differ from other species in that the males assist the females with the care of the young.

Of all the "aqua-shows" performed by birds whose natural habitat is the water, none are more intricate and fascinating than those of the grebes. These birds are an ancient family numbering only fourteen living species. They are distinguished from other swim-

ming birds by a number of anatomical peculiarities. To the observer, the two most immediately apparent of these features are the virtual lack of a tail and the position of the legs, which are set back so far on the body that the birds are all but helpless on dry land. This deficiency does not bother them at all, however, because these expert swimmers and divers never voluntarily leave the water, which is their exclusive habitat. Even the nest is built on the water; the floating structure is made of wet vegetation, and the eggs are often half-submerged—and always wet—during the period of incubation.

Grebes have a soft, lustrous, silky feather dress, for which they

A Great Crested Grebe

were hunted extensively in former years. Both sexes of several species grow a breeding plumage that may include conspicuous "horns" or "ears," thereby lending a distinctive appearance to the head.

Because of the bird's considerable size—it measures more than two feet in length—and its horned crest and feather collar, the courtship of the Great Crested Grebe is especially impressive. A common and widespread Eurasian species, this large grebe has long been a favorite object of study by ornithologists, resulting in detailed knowledge of its habits. In 1914, naturalist Sir Julian Huxley wrote a monograph about the courtship of the Great Crested Grebe which since has become a classic.

In order to get the attention of the female, the male grebe performs a dizzying dance in the water that includes alternate diving and rising to the surface in a jetlike thrust of watery spray close to the female. After she indicates her acceptance, the couple then perform simultaneously. One of the more spectacular phases of their ritual involves rising almost completely out of the water in an erect "penguin" position while facing one another, each with a piece of weed which they have brought up from the bottom dangling from its bill.

Grebes are monogamous and are faithful mates as well as good parents. Nest-building is an integral part of their courtship, and is shared by both partners, as are the duties of incubating the eggs and later feeding, teaching, and protecting the offspring. Young grebes are not born with the expert diving abilities of the adults; the parents, usually the mothers, must teach them the fine art of diving as well as the skill of catching fish and other aquatic creatures that make up the grebes' diet. Should the female lay another clutch of eggs before the young of the first brood are ready to fend for themselves, the father takes charge of the youngsters while his mate incubates the second batch of eggs.

In contrast to waterfowl, the majority of aquatic birds such as gulls and terns that are found along the seashores are colony breeders, whose nesting sites are often so crowded together that there is almost constant bickering among neighbors trespassing on each other's small territories. The habits of some species—the common North American Laughing Gull, for instance—have been studied in detail. Among other things, these observations have revealed the elaborate courtship ritual of the gulls, in which each sex has its own distinctive part to play.

Since male and female look exactly alike, not only to the human observer but also to the gulls themselves, the first move in the

courtship game is the male's attempt to locate an individual of the opposite sex. This he does by charging a likely-looking gull to see how the bird reacts. If the charge is met by a countercharge, the individual is a male, and the disappointed suitor has to look elsewhere. Should the threatened gull remain standing rigidly erect, however, without making a move except to turn its head sideways, the male has been successful in locating a prospective bride, and the mutual performance can begin.

Assuming a stiffly erect posture, the couple face in opposite directions and slowly turn their heads from side to side in a ceremony that evidently seals the pairing off of the two individuals. The courtship performance is not complete, however, until the symbolic food-begging ritual takes place. As mentioned earlier, actual or symbolical ritual feeding is a part of many avian courtship displays, and may continue throughout the entire breeding season. Some biologists believe that the females of some species may need an increased caloric intake at that time. Among terns, which are

close relatives of the gulls but look more swallowlike in flight, offerings of food occur regularly during courtship, most often apparently shortly before mating. The male Common Tern has been observed bringing back a minnow and then facing his mate, wings lowered and head pointing skyward, with the fish dangling from his bill, while she crouched and begged for the food. The courtship ritual, which in many ways is similar to that of the gulls, includes much posturing with stiffly erect tails and drooping wings.

Among Laughing Gulls, the feeding ritual is symbolical. Although both partners make motions of begging for food, the postures of the sexes differ considerably; the crouching, subordinate stance of the female is a prerequisite for mating because it establishes the dominance of the male. Soon afterward, the pair leaves the communal courtship area for the breeding site, where each couple claims its own small nesting territory.

Probably the most impressive among the prenuptial performances of marine birds is that of the Wandering Albatross, even if that ceremony were less bizarre than it is, the performers' large size and eleven-foot wingspread—the largest of any bird—would suffice to make it a spectacular show. In this ritual, the partners face one another and engage in a grotesque, awkward dance in which they bow and scrape, spread their huge wings, and point their bills skyward while emitting the most peculiar groans and clapping noises made by snapping their bills. Several other individuals may join in the performance before the pairing off is definitely settled.

Like most other oceanic birds, albatrosses are colony nesters, usually preferring offshore islands, where each couple select a site and fashion a crude, grass-lined nest of mud and soil. The pair take turns in incubating their single egg; the incubation period for the Wandering Albatross, largest of the thirteen species in this group, is about eleven weeks. The growth period of the young is correspondingly long, which explains why these birds breed only once every two years. Sexual maturity likewise comes very late for the albatross; there is reason to assume that the large species may not be ready to mate until they are at least eight years old.

To a greater or lesser degree, all the instances of avian courtship display—from the birds of paradise to the albatross—described and depicted in the preceding pages involve some kind of strange language, or code, which may be visual, vocal, or both. Although obviously meaningful to the birds, this code must remain at least partly alien and unintelligible to humans. On the other hand, there also is much that strikes a responsive chord in any human observer. Biologists warn against the temptation to assign our own feelings and reasonings to animals; these warnings are generally justified, yet it is difficult—and perhaps not even altogether desirable—to refrain from drawing at least some parallels. Such comparisons are, after all, less an attempt to read human traits and characteristics into animals than recognition of the fact that we have retained many of theirs in our nature, and that these traits are by no means always the manifestation of coarse, callous sensuality, especially in sexual matters, that most people think of when they state that a human being behaves "like an animal."

A case in point, and perhaps a fitting end to this parade of avian courtship behavior, which began with the showiest and most exotic birds, is that of the common Eurasian Jackdaw. This small relative of the crows lacks any special plumes, bright colors, or attractive feather patterns, but its habits are well worth a close look.

One of the best-qualified and most-experienced observers of these birds is Konrad Lorenz, mentioned earlier in connection with another matter involving Jackdaws. The German naturalist for years gathered a wealth of detailed data about the habits and the behavior of these fascinating birds, flocks of which made their home in the attic of his house year after year, and which he describes in his book *King Solomon's Ring*.

Despite the fact that Jackdaws are highly social birds that aggregate in flocks, roost together, and nest in colonies, they are monogamous and, once paired off, remain faithful to their mates for life. They also have a complicated social caste, or rank, system in which each bird has its own place and knows it. The dominant individual occupies the top rung on this social ladder, and the lowest-ranking bird has to be satisfied with the bottom rung.

Jackdaws form attachments to individuals of the opposite sex in the year before they become sexually mature. At such time, a single male will start looking around for a single female. When he finds one he likes, he immediately begins to court her. Lorenz found the suitors' subsequent shenanigans highly amusing, and in many ways reminiscent of human behavior in similar circumstances. Making every effort to impress his prospective bride favorably, the male stalks about with his head held high, and occasionally picks a fight with another male—but only if "she" is watching! By ogling her constantly, he makes it quite clear that he cannot take his eyes off her. Should the female be interested, her behavior is no less amusing, for she puts on a fine act of feigned indifference, pretending to look anywhere and everywhere except at her suitor. In reality, however, she does throw him a quick glance now and then, which is sufficient encouragement for the male to continue with his courtship. Should she genuinely ignore him, he would soon take the hint and give up his hopeless quest.

The female eventually signals her acceptance by cowering down before the stiffly erect male, her wings and tail trembling. Although this probably is a symbolic mating posture, it presently signifies nothing more than a submissive ceremonial greeting, used by females of this species throughout their lives, and at all times of the year, to welcome back their mates after a temporary absence.

As soon as she is firmly "engaged," a great deal changes for the female. In accordance with an apparently more or less inflexible rule of the Jackdaw social system, a male can choose his mate only from among the lower-ranking females; the one he selects as his "fiancée" then moves up to his own rank. In this way, a very low-ranking female chosen by the dominant male can become "top bird." Sad to relate, such females tend to be rather haughty and behave disdainfully toward their former companions of the lower ranks. Here, Jackdaws display traits that appear all-too human!

For one another, however, the newly paired-off couple have nothing but concern and tenderness. They are together constantly. Should the male find a particularly tasty tidbit, he brings it to the female, who accepts it in the food-begging posture of a fledgling bird. She, on the other hand, appears ever attentive and solicitous,

frequently expressing this by gently but thoroughly preening the feathers at the back of his head, a procedure he seems to find especially pleasurable.

The most unusual and touching fact about this behavior is its constancy; the "honeymoon" that begins even before sexual maturity apparently lasts for life. A Jackdaw couple live, love, feed, and fight off attacks together, raise their young together, and throughout it all remain faithful and attentive to one another. For a bird —or anyone else—that is quite a record.

Spanning the entire range from the spectacular and impressive to the graceful and tender, from the ornate and beautiful to the bizarre and comical, and displaying every variation of color, pattern, shape, sound, movement, and behavior, the courtship of birds is, among other things, a testimonial to the overwhelming urge for diversity in nature, life's infinite variety that makes it so infinitely fascinating.

Scientific Names of Birds Illustrated in This Book

(Listed in Order of Appearance)

CHAPTER 1

American Robin, *Turdus migratorius*, p. 14
Red Jungle Fowl, *Gallus gallus*, p. 17
Canada Goose, *Branta canadensis*, p. 18
Lesser Whitethroat, *Sylvia curruca*, p. 21
Chaffinch, *Fringilla fringilla*, p. 21
Common Peacock, *Pavo cristatus* (albino form), pp. 24–25
Nightingale, *Luscinia megarhynchos*, p. 30
White Bellbird, *Procnias alba*, p. 33
Frilled Coquette Hummingbird, *Lophornis magnifica*, p. 34
Pennant-winged Nightjar, *Semeiophorus vexillarius*, p. 37

CHAPTER 2

Greater Bird of Paradise, *Paradisea apoda*, pp. 38, 44–45
Blue Bird of Paradise, *Paradisea rudolphi*, pp. 48–49
Magnificent Bird of Paradise, *Diphyllodes magnificus*, pp. 50–51
Superb (Common) Lyrebird, *Menura novaehollandiae*,
pp. 54–55, 56–57, 58–59
Great Argus Pheasant, *Argusianus argus*, pp. 60–61, 64–65
Common Peacock, *Pavo cristatus*, pp. 66–67, 68–69, 72–73

CHAPTER 3

Black Grouse, *Lyrurus tetrix,* p. 74
Capercaillie, *Tetrao urogallus,* pp. 78–79
Sage Grouse, *Centrocercus urophasianus,* pp. 83, 84–85
Prairie Chicken, *Tympanuchus cupido,* pp. 82, 86–87
Ruff, *Philomachus pugnax,* pp. 90–91
Great Bustard, *Otis tarda,* pp. 94–95, 96–97
African Ostrich, *Struthio camelus,* pp. 100–101
Red-headed (Yellow-thighed) Manakin, *Pipra mentalis,* pp. 102–103
Paradise Widowbird, *Vidua paradisea,* pp. 106–107

CHAPTER 4

Crestless Gardener Bowerbird, *Amblyornis inornatus,* pp. 108, 114–115
Great Gray Bowerbird, *Chlamydera nuchalis,* pp. 118–119
Regent Bowerbird, *Sericulus chrysocephalus,* pp. 120–121
Satin Bowerbird, *Ptilonorhynchus violaceus,* pp. 124–125, 126–127
Magnificent Frigate Bird, *Fregata magnifica,* p. 130
Adélie Penguin, *Pigoscelis adeliae,* pp. 132–133
Penduline Titmouse, *Remiz pendulinus,* pp. 136–137
Winter Wren, *Troglodytes troglodytes,* pp. 138–139
Village (Textor) Weaver, *Ploceus cucullatus,* pp. 142–143

CHAPTER 5

Crowned Crane, *Balearica pavonina,* p. 146
Whooping Crane, *Grus americana,* pp. 150–151, 152–153, 154–155
Trumpeter Swan, *Cygnus buccinator,* pp. 158–159, 160–161
Ruddy Duck, *Oxyura jamaicensis,* pp. 164–165
Great Crested Grebe, *Podiceps cristatus,* pp. 166, 168–169
Laughing Gull, *Larus atricilla,* pp. 170–171, 172–173, 174–175
Wandering Albatross, *Diomedea exulans,* pp. 176–177
Jackdaw, *Corvus monedula,* pp. 180–181

Bibliography

Armstrong, E. A. 1942. Bird Display: an introduction to the study of bird psychology. Cambridge University Press

Beebe, W. 1936. Pheasants, their lives and homes. New York, Doubleday

Gilliard, E. T. 1969. Birds of Paradise and Bowerbirds. Garden City, Natural History Press

__________1958. Feathered Dancers of Little Tobago. National Geographic Magazine, Vol. 114

__________1958. Living Birds of the World. New York, Doubleday

Iredale, T. 1956. Birds of New Guinea. Melbourne, Georgian House

Knowlton, F. H. 1909. Birds of the World. New York, Holt

Lorenz, K. Z. 1952. King Solomon's Ring. New York, Crowell

Portmann, A. 1965. Aufbruch der Lebensforschung. Zuerich, Rhein-Verlag

__________1965. Neue Wege der Biologie. Munich, R. Piper & Co.

__________1957. Von Voegeln und Insekten. Basel, Friedrich Reinhardt A. G.

Von Frisch, K. 1974. Animal Architecture. New York, Harcourt Brace Jovanovich

Index

Pages on which illustrations appear are shown in *italics*.